Studies in Computational Intelligence 456

For further volumes:
http://www.springer.com/series/7092

Ireneusz Czarnowski, Piotr Jędrzejowicz,
and Janusz Kacprzyk (Eds.)

Agent-Based Optimization

Editors
Dr. Ireneusz Czarnowski
Gdynia Maritime University
Gdynia
Poland

Prof. Piotr Jędrzejowicz
Gdynia Maritime University
Gdynia
Poland

Prof. Janusz Kacprzyk
Systems Research Institute
Polish Academy of Science
Warszawa
Poland

and

PIAP-Industrial Institute of Automation
and Measurements
Warszawa
Poland

ISSN 1860-949X
e-ISSN 1860-9503
ISBN 978-3-642-44731-0
ISBN 978-3-642-34097-0 (eBook)
DOI 10.1007/978-3-642-34097-0
Springer Heidelberg New York Dordrecht London

Printed on acid-free paper

Springer is part of Springer Science+Business Media (www.springer.com)

Preface

Agent based systems technology has emerged as a new paradigm for the conceptualization, design, analysis and implementation of many approaches and solutions, notably in the area of software systems. Basically, software agents are sophisticated pieces of software that are meant to autonomously, on behalf of their users, solve a variety of complex tasks and problems. An important characteristic feature is that the agents are primarily meant to operate in and across open and distributed
environments.

A natural implication of a growing complexity of problems to be solved by agents is that multiple agents are needed that can work together. The concept of a *multi-agent system* has been advocated in this respect as an effective and efficient solution. Basically, the multi-agent system is a loosely coupled network of software agents that interact in an autonomous, collaborative and possibly synergistic way to solve problems that are beyond the individual capacities of a single problem solving agent.

It is easy to see that such a multi-agent architecture can have many advantages over a simplistic single agent approach. To name a few, one quote in this context the following ones. A multi-agent systems distributes problem solving and computational resources and capabilities across a network of interconnected agents. Since it is decentralized and distributed, various problems related to resource and performance limitations and bottlenecks, disastrous failures, etc. can be alleviated to a large extent, somehow even overcome. The setting in which problems are modeled using the multi-agent systems paradigm is in terms of autonomous interacting components which, as research results from many fields suggest, is a natural and more effective and efficient way of representing and solving many problems, notably related to the broadly perceived rational decision making in large, and spatially and temporally distributed systems. Moreover, the multi-agent systems often exhibit an enhanced performance with respect to computational efficiency, reliability, extensibility, robustness, maintainability, responsiveness, flexibility, and reuse.

Broadly perceived rational decision making is an omnipresent meta-problem in both science and technology, and even everyday life, because all kinds of human activities, and also their analogues in inanimate systems, call for taking most advantage of what is intended and possible to attain. Examples are here innumerable and range from best individual investment decisions, through all kinds of best design solutions for products and services to best fulfill societal needs and expectations.

Since it is generally assumed that the solutions of the above mentioned problems of rational choice is best performed in terms of formal problem formulations and analyses, decision making and decision analytic models immediately become a natural choice. Among them a special role is played by optimization and mathematical programming models in which the problem is naturally set in terms of the maximization or minimization of some objective (performance) function under some constraints conveniently given as a set of equalities and/or inequalities. Needless to say that such a setting is a clear reflection of quite a natural utility maximization type of rationality.

This volume is a collection of original research works by leading specialists focusing on novel and promising approaches in which the multi-agent system paradigm is used to support, enhance or replace traditional approaches to solving difficult optimization problems. The editors have invited several well-known specialists to present their solutions, tools, and models falling under the common denominator of the *agent-based optimization*. The book consists of eight chapters covering examples of application of the multi-agent paradigm and respective customized tools to solve difficult optimization problems arising in different areas such as machine learning, scheduling, transportation and, more generally, distributed and cooperative problem solving.

The chapter of Ireneusz Czarnowski and Piotr Jędrzejowicz ("Machine Learning and Multiagent Systems as Interrelated Technologies") contains a short, yet comprehensive account of main applications in which machine learning methods have been used to support agent learning capabilities. Then, current research results integrating machine learning and agent technologies are reviewed in a more detailed manner. Examples of machine learning models proposed by the authors in which the agent paradigm has been implemented, aimed at finding optimal solutions to machine learning tasks, are included in the main part of the chapter. A general conclusion and message of the chapter is that agent technologies can effectively support finding good quality solutions to machine learning problems.

Mariusz Boryczka and Wojciech Bura ("Ant Colony Optimization for the Multicriteria Vehicle Navigation Problem") describe a family of multi-agent ant-based vehicle navigator algorithms dedicated to the solution of a multi-criterion problem of finding the shortest path between two points on the map. Sequential and parallel versions of the multi-agent ant-based algorithm are presented. The work shows that the parallel versions of the algorithm can be implemented using the graphics processing unit (GPU) and the CUDA architecture.

The chapter written by Dariusz Barbucha ("Solving Instances of the Capacitated Vehicle Routing Problem Using Multi-Agent Non-Distributed and Distributed Environment") focuses on the design and use of the asynchronous team of agents (A-Team) implemented using the JADE-Based A-Team middleware environment (JABAT). The JABAT multi-agent system has been designed for solving computationally hard optimization problems using the parallel and distributed environment.

JABAT produces solutions to combinatorial optimization problems using a set of optimization agents, each representing an improvement algorithm. The chapter focuses on the ability of JABAT to distribute computation load while system is engaged in solving instances of some difficult optimization problem. The experiments reported are carried out on instances of the capacitated vehicle routing problem.

Piotr Jędrzejowicz and Aleksander Skakovski ("Structure vs. Efficiency of the Cross-Entropy Based Population Learning Algorithm for Discrete-Continuous Scheduling with Continuous Resource Discretisation") deal with the application of the population learning algorithm for solving discrete-continuous scheduling problem with the continuous resource discretization. The population learning algorithm evolves primary solutions delivered by the cross-entropy method. Different versions of the population learning algorithm, viewed as an island model, that differ from each other by their structure and migration scheme, are discussed and evaluated experimentally. In the approach considered each island can be treated as an independent agent cooperating with other agents. The authors provide and answer to the question whether the topology of learning stages (or islands) might have some effect on the efficiency of the algorithm.

Another scheduling problem is considered in the chapter by Piotr Jędrzejowicz and Ewa Ratajczak-Ropel ("Triple-Action Agents Solving the MRCPSP/max Problem"). The authors propose an A-Team architecture for solving the multi-mode resource-constrained project scheduling problem with the minimal and maximal time lags. A computational experiment involves the performance evaluation of optimization agents within the A-Team proposed. The results obtained show that the A-Team implementation is an effective tool for solving instances of the scheduling problem considered.

Effects and impact of cooperation between the cooperating A-Teams working parallel and combined into an architecture designed for solving difficult combinatorial optimization problems is discussed in the chapter written by Dariusz Barbuch, Ireneusz Czarnowski, Piotr Jędrzejowicz, Ewa Ratajczak-Ropel and Izabela Wierzbowska ("Team of A-Teams - a Study of the Cooperation Between Program Agents Solving Difficult Optimization Problems"). Several architectures of cooperation are compared. Experimental results show that in most cases the integration of the distributed evolutionary concept, and especially the island based evolutionary algorithm, with the A-Team paradigm might result in a noticeable improvement of the quality of the computation results. In conclusion the authors confirm the importance of choosing an effective and efficient information exchange architecture for the team of agents solving a particular optimization task.

Kunal Srivastava, Angelia Nedić and Dušan Stipanović ("Distributed Bregman-Distance Algorithms for Min-Max Optimization") focuse on a min-max optimization problem over a time-varying network of computational agents. To solve the problem the authors propose distributed subgradient algorithms in which each agent computes its own estimates of an optimal point based on its own cost function, and then communicates these estimates to its neighbors in the network. Two algorithms are discussed, both using the Bregman-distance functions. The applicability of the algorithms is demonstrated by considering a power allocation problem in a cellular network.

In the last chapter, Anand J. Kulkarni and Kang Tai ("A Probability Collectives Approach for Multi-Agent Distributed and Cooperative Optimization with Tolerance for Agent Failure") consider an inherent ability of the distributed and decentralized agent-based optimization technique, referred to as Probability Collectives (PC), to accommodate agent failures. It is assumed that complex systems can be decomposed into smaller subsystems to be further treated in a distributed and decentralized way. The PC framework for the optimization of complex systems is presented. The ability of the PC approach to tolerate instances of agent failures based on solving of the circle packing problem is presented.

The editors strongly believe that this volume has been an important and timely initiative. The area of agent based optimization has become a mature field of research with many relevant analytic contribution and, what is particularly important, a rapidly increasing flow of implementations in many diverse areas which have clearly demonstrated the effectiveness and efficiency of the new approach. The editors have been privileged to collect a representative collection of relevant papers from leading researchers. They have provided the readers with an account of both deep and relevant formal and analytic results and an in depth presentation of real world applications. It is hoped that the presented ideas and results will be of value to the research community working in the field of artificial intelligence, knowledge discovery, collective computational intelligence, intelligent transportation systems, project management and, in particular, agent and multi-agent systems technologies and applications.

We would like to take this opportunity to thank all authors for their highly valuable contributions. We wish to thank all peer reviewers whose invaluable work and suggestions have helped improve the quality of the chapters. Special thanks are due to Professor Juan Antonio Rodriguez Aguilar, Dr. Xiafeng Li, Dr. Mahdi Zargayouna and Dr. Rafał Różycki for their multi-faceted help in the difficult process of preparation of this book. Dr. Tom Ditzinger and Dr. Leontina Di Cecco from Springer deserve our deep appreciation for a great collaboration and consideration.

Ireneusz Czarnowski
Piotr Jędrzejowicz
Janusz Kacprzyk

Contents

Machine Learning and Multiagent Systems as Interrelated Technologies 1
Ireneusz Czarnowski, Piotr Jędrzejowicz

Ant Colony Optimization for the Multi-criteria Vehicle Navigation Problem 29
Mariusz Boryczka, Wojciech Bura

Solving Instances of the Capacitated Vehicle Routing Problem Using Multi-agent Non-distributed and Distributed Environment 55
Dariusz Barbucha

Structure vs. Efficiency of the Cross-Entropy Based Population Learning Algorithm for Discrete-Continuous Scheduling with Continuous Resource Discretisation 77
Piotr Jędrzejowicz, Aleksander Skakovski

Triple-Action Agents Solving the MRCPSP/Max Problem 103
Piotr Jędrzejowicz, Ewa Ratajczak-Ropel

Team of A-Teams - A Study of the Cooperation between Program Agents Solving Difficult Optimization Problems 123
Dariusz Barbucha, Ireneusz Czarnowski, Piotr Jędrzejowicz, Ewa Ratajczak-Ropel, Izabela Wierzbowska

Distributed Bregman-Distance Algorithms for Min-Max Optimization 143
Kunal Srivastava, Angelia Nedić, Dušan Stipanović

A Probability Collectives Approach for Multi- agent Distributed and Cooperative Optimization with Tolerance for Agent Failure 175
Anand J. Kulkarni, Kang Tai

Author Index 203

Machine Learning and Multiagent Systems as Interrelated Technologies

Ireneusz Czarnowski* and Piotr Jędrzejowicz

Abstract. The chapter reviews current research results integrating machine learning and agent technologies. Although complementary solutions from both fields are discussed the focus is on using agent technology in the field of machine learning with a particular interest on applying agent-based solutions to supervised learning. The chapter contains a short review of applications, in which machine learning methods have been used to support agent learning capabilities. This is followed by a corresponding review of machine learning methods and tools in which agent technology plays an important role. Final part gives a more detailed description of some example machine learning models and solutions where the paradigm of the asynchronous team of agents has been implemented to support the machine learning methods, and which have been developed by the authors and their research group. It is argued that agent technology is particularly useful in case of dealing with the distributed machine learning problems. As an example of such applications a more detailed description of the agent-based framework for the consensus-based distributed data reduction is given in the final part of the chapter.

1 Introduction

Contemporary definition sees machine learning as a discipline that is concerned with the design and development of algorithms that allow computers to learn behaviors based on empirical data. Data can be seen as examples that illustrate relations between observed objects. A major focus of machine learning research is to automatically learn to recognize complex patterns and make intelligent decisions based on data.

Ireneusz Czarnowski · Piotr Jędrzejowicz
Department of Information Systems, Gdynia Maritime University,
Morska 83, 81-225 Gdynia, Poland
e-mail: {irek,pj}@am.gdynia.pl

* Corresponding author.

I. Czarnowski et al. (Eds.): Agent-Based Optimization, SCI 456, pp. 1–28.
DOI: 10.1007/978-3-642-34097-0_1

Parallel to recent developments in the field of machine learning, mainly as a result of convergence of many technologies within computer science such as object-oriented programming, distributed computing and artificial life, the agent technology has emerged. An agent is understood here as any piece of software that is designed to use intelligence to automatically carry out an assigned task, mainly retrieving and delivering information.

Tweedale and co-authors [80] outline an abridged history of agents as a guide for the reader to understand the trends and directions of future agent design. This description includes how agent technologies have developed using increasingly sophisticated techniques. It also indicates the transition of formal programming languages into object-oriented programming and how this transition facilitated a corresponding shift from scripted agents (bots) to agent-oriented designs which is best exemplified by multiple agent systems (MAS). According to Liau [49], a MAS tries to solve complex problems with entities called agents, using their collaborative and autonomous properties. Jennings et al. [44] list the following MAS properties:

- Each agent has partial information or limited capabilities.
- There is no global system control.
- Data in a MAS are decentralized.
- Computation is asynchronous.
- Different agents could be heterogeneous, for example, with respect to knowledge representation, reasoning model, solution evaluation criteria, goal, architecture or algorithm for task performance.

During the last decade developments in the fields of machine learning and agent technologies have, in some respect, become complementary and researchers from both fields have seen ample opportunities to profit from solutions proposed by each other. Several agent-based frameworks that utilize machine learning for intelligent decision support have been recently reported. Learning is increasingly being seen as a key ability of agents, and researchinto learning agent technology, such as reinforcement learning and supervised or unsupervised learning has produced many valuable applications.

In this chapter the focus is however on using agent technology in the field of machine learning with a particular interest on applying agent-based solutions to supervised learning. Supervised learning is the machine learning task of inducing a function from training data which is a set of training examples. In supervised learning, each example is a pair consisting of an input object (typically a vector) and a desired output value (also called the supervisory signal or class label). A supervised learning algorithm analyzes the training data and produces an induced function, which is called a classifier if the output is discrete, or a regression function if the output is continuous. The inferred function should predict the correct output value for any valid input object. This requires the learning algorithm to generalize from the training data to unseen situations.

There are several ways the machine learning algorithm can profit from applying agent technology. Among them the following will be addressed in this paper:

- There are numerous machine learning techniques where parallelization can speed-up or even enable learning. Using a set of agents may, in such circumstances, increase efficiency of learning.
- Several machine learning techniques directly rely on the collective computational intelligence paradigm, where a synergetic effect is expected from combining efforts of various program agents.
- There is a class of machine learning problems known as the distributed machine learning. In the distributed learning a set of agents working in the distributed sites can be used to produce some local level solutions independently and in parallel. Later on local level solutions are combined into a global solution.

The chapter is organized as follows. Section 2 contains a short review of applications, in which machine learning methods have been used to support agent learning capabilities. Section 3 offers a corresponding review of machine learning methods and tools in which agent technology plays an important role. Section 4 gives more detailed description of some example machine learning models and solutions where the agent paradigm has been implemented and which have been developed by the author and his research group. Finally, conclusions contain suggestions on future research and possible deeper integration of machine learning and agent technology.

Content of the chapter integrates and extends two earlier papers of Czarnowski and Jędrzejowicz([41] and [27]) presented, respectively, at the KES-AMSTA Conference, Manchester 2011 and IEEE System, Man, and Cybernetics Conference, Anchorage 2011.

2 Learning Agents

Probably the most often used approach to provide agents with learning capabilities is the reinforcement learning. An excellent survey of multiagent reinforcement learning can be found in the paper of Busoniu et al. [13]. As it was pointed out by Sutton and Barto [74] reinforcement learning is learning what to do - how to map situations to actions - so as to maximize a numerical reward signal. The learner is not told which actions to take, as in most forms of machine learning, but instead must discover which actions yield the most reward. In describing properties of the reinforcement learning the authors directly refer to the notion of agent. In their view a learning agent must be able to sense the state of the environment and must be able to take actions that affect the state. The agent also must have goal or goals relating to the state of the environment [74].

Theoretical developments in the field of learning agents focus mostly on methodologies and requirements for constructing multiagent systems with learning capabilities. Connection of the theory of automata with the multiagent reinforcement learning is explored in Nowe et al. [58]. Shoham et al. [68] claim that the area of learning in multiagent systems is today one of the most fertile grounds for interaction between game theory and artificial intelligence. In the paper by Mannor and Shamma [52]challenges motivated by engineering applications and the potential appeal of multi-agent learning to meetthese challenges are discussed.

Symeonidis et al. [75] present an approach that takes the relevant limitations and considerations into account and provides a gateway on the way data mining techniques can be employed in order to augment agent intelligence. This work demonstrates how the extracted knowledge can be used for the formulation initially, and the improvement, in the long run, of agent reasoning. Preux et al. [61] present MAABAC, a generic model for building adaptive agents: they learn new behaviors by interacting with their environment. Agents adapt their behavior by way of reinforcement learning, namely temporal difference methods. The paper by Sardinhaet et al. [67] presents a systematic approach to introduce machine learning in the design and implementation phases of a software agent. It also presents an incremental implementation process for building asynchronous and distributed agents, which supports the combination of machine learning strategies. Rosaci in [66] proposes a complete MAS architecture, called connectionist learning and inter-ontology similarities (CILIOS), for supporting agent mutual monitoring.

In the paper of Masoumi and Meybodi [53] the concepts of stigmetry and entropy are imported into learning automata based multi-agent systems with the purpose of providing a simple framework for interaction and coordination in multi-agent systems and speeding up the learning process. Another extension was proposed by Boylu et al. [11]. The authors suggest a merging, and hence an extension, of two recent learning methods, utility-based learning and strategic or adversarial learning. Utility-based learning brings to the forefront the learner's utility function during induction. Strategic learning anticipates strategic activity in the induction process when the instances are intelligent agents such as in classification problems involving people or organizations. The resulting merged model is called the principal-agent learning. Loizos [50] argues that when sensing its environment, an agent often receives information that only partially describes the current state of affairs. The agent then attempts to predict what it has not sensed, by using other pieces of information available through its sensors. Machine learning techniques can naturally aid this task, by providing the agent with the rules to be used for making these predictions. For this to happen, however, learning algorithms need to be developed that can deal with missing information in the learning examples in a principled manner, and without the need for external supervision. It is shown that the *Probably Approximately Correct* semantics can be extended to deal with missing information during both the learning and the evaluation phase.

Numerous reinforcement learning applications have been recently reported in the literature. Some interesting examples include a proposal of reinforcement learning for agent-based production scheduling proposed by Wang and Usher [81], and a case-based reinforcement learning algorithm (CRL) for dynamic inventory control in a multi-agent supply-chain system of Jiang and Sheng [45]. A comprehensive survey of multiagent reinforcement learning algorithms, solutions and approaches is offered in [13]. A reinforcement learning (RL) agent learns by trial-and-error interaction with its dynamic environment. At each time step, the agent perceives the complete state of the environment and takes an action, which causes the environment to transit into a new state. The agent receives a scalar reward signal that evaluates the quality of this transition. Well understood and effective algorithms are available for solving single agent reinforced learning. Much difficult problems are posed by the multiagent learning case. According to Busoniu et al. [13] main techniques used to solve multiagent reinforcement learning include temporal difference RL, game theory and direct policy search. Recent example of the approach based on the first of the above mentioned techniques is a multi-goal Q-learning algorithm of cooperative teams proposed by Li et al. [48].

Supervised learning techniques have been also applied to support agent's learning capabilities. In the paper of Yu et al. [83], a support vector machine (SVM) based multiagent ensemble learning approach is proposed for credit risk evaluation. Different SVM learning paradigms with much dissimilarity are constructed as intelligent agents for credit risk evaluation. Multiple individual SVM agents are trained using training subsets. In the final stage, all individual results produced by multiple SVM agents in the previous stage are aggregated into an ensemble result.

An interesting example of integration of the agent and machine learning technologies was proposed in [76]. The above authors have developed Agent Academy, an integrated development framework that supports both design and control of multi-agent systems (MAS), as well as "agent training". They define agent training as the automated incorporation of logic structures generated through data mining into the agents of the system. The increased flexibility and cooperation primitives of MAS, augmented with the training and retraining capabilities of Agent Academy, provide a powerful means for the dynamic exploitation of data mining extracted knowledge. In their paper, the methodology and tools for agent retraining are presented. Through experimented results with the Agent Academy platform, it was demonstrated how the extracted knowledge can be formulated and how retraining can lead to the improvement - in the long run - of agent intelligence.

3 Agent-Based Machine Learning

Recently, several machine learning solutions and techniques have been reported to rely on applying agent technologies. Such solutions and techniques belong to the two broad classes - universal ones and dedicated to particular applications. Solutions and techniques belonging to the first class involve applications of the multi agent systems, including A-Teams and the population-based methods. This section contains a review of some recent universal and dedicated solutions with the exception of those based on the A-Team paradigm. Machine learning solutions using the A-Team paradigm are discussed in a detailed manner in Section 4.

3.1 Universal Solutions and Techniques

As it has been observed by Luo et al. [51] industry, science, and commerce fields often need to analyze very large datasets maintained over geographically distributed sites by using the computational power of distributed systems. The Grid can play a significant role in providing an effective computational infrastructure support for this kind of data mining. Similarly, the advent of multi-agent systems has brought us a new paradigm for the development of complex distributed applications.Through a combination of these two techniques an Agent Grid Intelligent Platform and an integrated toolkit VAStudio used as a testbed were proposed. Using grid platform as a testbed was also suggested by Raicevic [65]. The author presents a parallel learning method for agents with an actor-critic architecture based on artificial neural networks. The agents have multiple modules, where the modules can learn in parallel to further increase learning speed. Each module solves a sub-problem and receives its own separate reward signal with all modules trained concurrently. The method is used on a grid world navigation task showing that parallel learning can significantly reduce learning time.

Kitakoshi et al. [46] describe an on-line reinforcement learning system that adapts to environmental changes using a mixture of Bayesian networks. Machine learning approaches, such as those using reinforcement learning methods and stochastic models, have been used to acquire behavior appropriate to environments characterized by uncertainty. The results of several experiments demonstrated that an agent using the proposed system can flexibly adapt to various kinds of environmental changes.

Gifford in his Ph.D. dissertation [32] advocates an approach focused on the effects of sharing knowledge and collaboration of multiple heterogeneous, intelligent agents (hardware or software) which work together to learn a task. As each agent employs a different machine learning technique, the system consists of multiple knowledge sources and their respective heterogeneous knowledge representations. Experiments have been performed that vary the team composition in terms of machine learning algorithms and learning strategies employed by the agents. General findings from these experiments suggest that constructing a team of classifiers using a heterogeneous mixture of homogeneous teams is preferred.

Quteish et al. at [64] proposed a neural network (NN)-based multi-agent classifier system (MACS) using the trust, negotiation, and communication (TNC) reasoning model. The main contribution of this work is that a novel trust measurement method, based on the recognition and rejection rates, was suggested. Two agent teams are formed; each consists of three NN learning agents. The first is a fuzzy min-max (FMM) NN agent team and the second is an ARTMAP a fuzzy adaptive resonance theory map (FAM) NN agent team. Modifications to the FMM and FAM models are proposed so that they can be used for trust measurement in the TNC model.

Several important methods can be grouped under the umbrella of the collective or collaborative learning. In the paper by Hoenl and Tuyls [35] it was shown show how Evolutionary Dynamics (ED) can be used as a model for Q-learning in stochastic games. Analysis of the evolutionary stable strategies and attractors of the derived

ED from the Reinforcement Learning (RL) application then predict the desired parameters for RL in multiagent systems to achieve Nash equilibriums with high utility. Secondly, it was shown how the derived fine tuning of parameter settings from the ED can support application of the COllective INtelligence (COIN) framework. COIN is a proved engineering approach for learning of cooperative tasks in MASs. In their paper Hofmann and Basilico [34] propose a collaborative machine learning framework to exploit inter-user similarities. More specifically, they present a kernel-based learning architecture that generalizes the well-known Support Vector Machine learning approach by enriching content descriptors with inter-user correlations.

Another umbrella covers learning classifier systems introduced by Holland [36] which use simple agents representing set of rules as a solution to a machine learning problem. A Pittsburgh-type LCS has a populations of separate rule sets, where the genetic algorithm recombines and reproduces the best of these rule sets. In a Michigan-style LCS there is only a single population and the algorithm's action focuses on selecting the best classifiers within that ruleset. Analysis of the properties of LCSs, comparison of several proposed variants and overview of the state of the art can be found in the papers [6], [7], [12] and [82]. Useful extension of the LCS concept was proposed by Smith et al. [72]. Their paper introduces a new variety of learning classifier system (LCS), called MILCS, which utilizes mutual information as fitness feedback. Unlike most LCSs, MILCS is specifically designed for supervised learning. Yet another extension introduces a mechanism for recognizing a current situation by determining a boundary between self and others, and investigates its capability through interaction with an agent [77]. An integration of several cognitively inspired anticipation and anticipatory learning mechanisms in an autonomous agent architecture, the Learning Intelligent Distribution Agent (LIDA) system was proposed by Negatu et al. [55].

Ensemble techniques have proven to be very successful in boosting the performance of several types of machine learning methods. In the paper by Bacardit and Krasnogor [8] authors illustrate usefulness of the ensemble techniques in combination with GAssist, a Pittsburgh-style Learning Classifier System. Effective and competitive ensembles constructed from simple agents represented by expression trees induced using Gene Expression Programming have been proposed in papers of Jędrzejowicz and Jędrzejowicz [42] and [43]. Their approach has been tested using several ensemble constructing techniques including AdaBoost learning, voting pool of classifiers, incremental learning, cluster based learning, mass functions based learning and meta-learning.

An effective approach to machine learning is to use some simple agents cooperating directly or indirectly during the process of learning. For example, Hong et al. [37] attempt to propose an Ant Colony System-based framework for fuzzy data mining. Their idea is to use simple agents (ants) to mine various types of data. In the framework, the membership functions are first encoded into binary-bits and then fed into the ACS to search for the optimal set of membership functions. The problem is then transformed into a multi-stage graph, with each route representing a possible set of membership functions. When the termination condition is reached, the best membership function set (with the highest fitness value) can then be used to mine fuzzy association rules from a database. Some earlier works on ACS-based rule discovery were proposed by Parpinelli et al. [59] as well as Cordon and Herrera [14], which proposed the mining of classification rules for fuzzy control systems.

Agent technology seems to be a natural tool for the distributed systems. Combining approaches to distributed learning with agent technology is considered as the promising and at the same time challenging problem in the distributed learning research [47]. In the paper by Zhang et al. [85] an agent paradigm was proposed as a tool for integration of different techniques into an effective strategy of learning from data. The proposed hybrid learning system integrates basic components of the learning process. Data pre-processing, selection, transformation and induction of the learning and post-learning models are carried out by a set of agents cooperating during the task execution. Several agent-based architectures have already been proposed to solve the distributed learning problems. It is usually assumed that each site can have one or more associated agents, processing the local data and communicating the results to other agents that control and manage the knowledge discovery process. Examples include Papyrus [62], MALE [69], ANIMALS [73], and MALEF [79]. In the paper of Albashiri et al. [3] EMADS, a hybrid peer-to-peer agent based system comprising a collection of the collaborating agents distributed across a network, was described.

According to Zhu et al. [86] data mining from distributed sources focuses on:

- Identifying locally significant patterns in individual databases
- Discovering emerging significant patterns after unifying distributed databases in a single view
- Finding patterns which follow special relationships across different data collections.

To solve the third problem, the above authors advocate a cross-database pruning concept and propose a collaborative pattern (CLAP) mining framework with cross-database pruning mechanisms for distributed pattern mining. In CLAP, distributed databases collaboratively exchange pattern information between sites so that each site can leverage information from other sites to gain cross-database pruning. The proposed framework with distributed sites playing agents roles, allows to carry out mining activities in a cooperative manner. CLAP allows the distributed sites to communicate with each other and exchange messages, so the mining is carried out at distributed sites without any data integration.

Similar views on close interrelations between agent and machine learning technologies have been expressed by several other specialists. For example, da Silva et al. [70] observed in their study of distributed data mining and agents that there exists a synergy between multiple agent systems and distributed data mining technology. The above paper provides adetailed literature review of existing distributed clustering algorithms (including privacy-preserving ones). In [70] it was also observed that scalable analysis of data may require advanced data mining for detecting hidden patterns, constructing predictive models, and identifying outliers, among others. In a MAS this knowledge is usually collective. This collective "intelligence" of a MAS must be developed by distributed domain knowledge and analysis of distributed data observed by different agents. Such distributed data analysis may be a non-trivial problem when the underlying task is not completely decomposable and computing resources are constrained by several factors.

3.2 Dedicated Solutions and Techniques

In the machine learning literature numerous applications solving particular machine learning problem type or task where agent technology have played an important, even if supporting, role have been recently reported. In this short review the focus is on some example cases where agent technology has been used in an innovative manner.

Fan et al. [31] have developed a two-stage model for personalized and intelligent information routing of online news. At the first stage, persistent user queries are extracted from rated documents based on Robertson's Selection Value (RSV). At the second stage, genetic programming is applied to discover the optimal ranking function for individual user. Pazzani and Billsus [60] developed a learning information agent called Syskill&Webert which could learn a user profile for the identification of interesting web documents. A separate user profile was created for each individual information topic. Web documents were represented as Boolean feature vectors, and each feature had a binary value indicating if a particular keyword appeared in the document or not. Feature selection was conducted based on Expected Information Gain which tends to select words appearing more frequently in positive documents. The classification mechanism of Syskill&Webertwas based on a naïve Bayesian classifier. The paper of Arevian et al. [4] focuses on symbolic transducers and recurrent neural preference machines to support the task of mining and classifying textual information. These encoding symbolic transducers and learning neural preference machines can be seen as independent agents, each one tackling the same task in a different manner.

Jansen [39] discusses various ways in which mobile agents could be applied to problem of detecting and responding to intrusions. Abraham et al. [1] proposed a distributed Intrusion Detection System (IDS) consisting of several IDS over a large network, all of which communicate with each other, or with a central server that facilitates advanced network monitoring. In a distributed environment, system is implemented using co-operative intelligent agents distributed across the network. To detect intrusions in a network three fuzzy rule based classifiers are constructed. Moskovitch et al. [54] conducted a comprehensive experiment for testing the feasibility of detecting unknown computer worms, employing several computer configurations, background applications, and user activities. During the experiments 323 computer features were monitored by the proposed agent. Four feature selection methods were used to reduce the number of features and four learning algorithms were applied on the resulting feature subsets.

In the paper of Jiang and Sheng [45] a case-based reinforcement learning algorithm (CRL) for dynamic inventory control in a multi-agent supply-chain system was proposed. In the paper of Gifford and Agah [33] authors utilize multi-agent machine learning and classifier combination to learn rock facies sequences from wire line well log data. The paper focuses on how to construct a successful set of classifiers, which periodically collaborate, to increase the classification accuracy. Utilizing multiple, heterogeneous collaborative learning agents is shown to be successful for this classification problem. Solving the pursuit problem with heterogeneous multiagent system using reinforcement learning was investigated by Ishiwaka [38].

Zhang and Zhang [84] present, a multiagent data warehousing (MADWH) and multiagent data mining (MADM) approach for brain modeling. An algorithm named Neighbor-Miner is proposed for MADWH and MADM. The algorithm is defined in an evolving dynamic environment with semiautonomous neurofuzzy agents. Instead of mining frequent itemsets from customer transactions, the new algorithm discovers new neurofuzzy agents and mines agent associations in first-order logic for coordination that was once considered impossible in traditional data mining.

One of the application areas where agent technology integrated with data mining solutions was used is e-learning. In the paper of Acampora et al. [2] the authors report an attempt to provide efficient and intelligent tools to be able to analyze learner's needs and preferences by exploiting an ontological representation of learning environment and an adaptive memetic approach, integrated into a cooperative multi-agent framework. In particular, a collection of agents analyzes learner preferences and generate high-quality learning presentations by executing, in a parallel way, different cooperating optimization strategies. This cooperation is performed by jointly exploiting data mining via fuzzy decision trees, together with a decision-making framework exploiting fuzzy methodologies.

Another example of the dedicated, agent-based data mining solution is a multi-agent framework for data mining in electromyography proposed by Balter et al. [9]. The aim is to extract medical information using data mining algorithms and to supply a knowledge base with pertinent information. The multi-agent platform gives the possibility to distribute the data management process between several autonomous entities. This framework provides a parallel and flexible data manipulation.

4 Machine Learning with A-Teams

Paradigms of the population-based methods and multiple agent systems have been during early nineties integrated within the concept of the asynchronous team of agents (A-Team). According to Talukdar et al. [78] an asynchronous team is a collection of software agents that cooperate to solve a problem by dynamically evolving a population of solutions. Current implementations of the A-Team concept are characterized by a high level of accessibility, scalability and portability. A review of the A-Team solutions and implementations can be found in the review by Jędrzejowicz [40]. All the machine learning solutions reviewed in this section are the A-Team implementations built using the JADE-Based A-Team middleware environment (JABAT) proposed by Barbucha et al. [10].

Two early applications of the A-Team paradigm to machine learning due to Czarnowski and Jędrzejowicz include training the cascade correlation learning architecture [19] and training the feed-forward artificial neural networks proposed in [20]. Recent solutions, described later on,are focused on data reduction and distributed learning.

Data reduction in the supervised machine learning aims at deciding which instances and which features from the original training set should be retained for further use during the learning process. Data reduction is considered as an important step towards increasing effectiveness of the learning process when the available training sets are large or distributed and when the access to data is limited and costly. Data reduction performed without losing extractable information can result in increased capabilities and generalization properties of the learning model. It is obvious that removing some instances from the training set reduces time and memory complexity of the learning process. The data reduction algorithms can be divided into two categories: prototype selection and prototype extraction. Prototype selection is a technique of choosing a subset of reference vectors from the original set, also by reduction of attributes, whereas prototype extraction means the construction of an entirely new set of instances, smaller, in respect to its dimensionality, than the original dataset. Prototype extraction can also include the process of feature construction, where decreasing the number of attributes is carried-out by creating new features onthe basis of some transformation of the original attributes. The performance criteria used in data reduction may include the accuracy of classification, the complexity of the hypothesis the classification costs and many other criteria.

The idea of applying agent technology to data reduction has been proposed in several papers of Czarnowski and Jędrzejowicz [16], [18], [21], [22], [25], [27]. In the above papers several architectures, models and strategies for the A-Team based data reduction have been proposed. Using them usually improves quality of the respective supervised machine learning. Most competitive results have been obtained by A-Teams producing clusters of instances from the training set and then selecting instances from these clusters.

Although a variety of methods could be used to produce clusters, using the similarity coefficient proposed in papers of Czarnowski and Jędrzejowicz [18] and [21]as the clustering criterion, have produced better than satisfactory results.

To solve the data reduction problem, several types of optimizing agents carrying out improvement procedures including tabu search, simulated annealing and variety of simple local search algorithms have been used. Basic assumptions behind the proposed approach are as follows:

- A solution is represented by a string consisting of two parts. The first contains numbers of instances selected as prototypes and the second - numbers of attributes chosen to represent the dataset.
- Prototype instances and attributes are selected from clusters through the population-based search carried out by the optimizing agents.
- Initially, potential solutions are generated by random selection of a single instance from each cluster and by random selection of the attribute numbers. Attributes are later adjusted by the attribute manager agent with a view to find the best combination and, at the same time, to unify the set of selected attributes at a global level (only in case of the distributed data reduction).

The solution manager is responsible for organizing the data reduction process through managing the population of solutions called individuals and updating them when appropriate. During the data reduction process the solution manager continues reading individuals (solutions) from the common memory and storing them back after attempted improvement until a stopping criterion is met. During this process the solution manager keeps sending randomly drawn individuals (solutions) from the common memory to optimizing agents. Each optimizing agent tries to improve the quality of the received solution and afterwards sends back the improved solution to the solution manager, which, in turn, updates common memory, replacing a randomly selected individual with the improved one. In each of the above cases the modified solution replaces the current one if it is evaluated as a better one. Evaluation of the solution is carried out by estimating classification accuracy of the classifier, which is constructed taking into account the instances and the attributes as indicated by the solution. Since the computational complexity of the above search procedures is linear, the computational complexity of the fitness evaluation is not greater than the complexity of the classifier induction. In case of the distributed data reduction an additional agent called attribute manager is used. Its role is to coordinate the attribute selection. The attribute manager agent is also responsible for the final integration of attributes selected locally by optimizing agents. The attribute manager actions include receiving candidate attributes form solution mangers, and deciding on the common set of attributes to be used at both the local and the global levels.

The idea of applying A-Team paradigm to solving the distributed learning problem has been evolving since a couple of years. Different solutions were presented in several papers of Czarnowski and Jędrzejowicz [15], [17], [23], [26] and [28]. The proposed approach, denoted as LCDD (Learning Classifiers from Distributed Data), involves two stages, both based on the collaboration between agents:

- Local, in which the selection of prototypes from the distributed data takes place (A-Teams are used to select prototypes by instance selection and/or removing irrelevant attributes).
- Global, consisting of pooling of the selected prototypes and producing the global learning model.

At the local level, that is, at the distributed data sources,agent-based population learning data reduction algorithms are executed in parallel. Instance and attribute reduction are integrated with the classifier learning process. An important feature of the LCDD approach is A-Teams ability to select instances and attributes in cooperation between agents, thus assuring a homogenous set of prototypes at the global level. In this case, the instance selection is carried out independently at each site through applying the agent-based population search but the attribute selection is managed and coordinated through the process of interaction and collaboration between agents. All the required steps of the proposed approach are carried out by program agents of the four following types:

- Global level manager - agent responsible for managing the process of the distributed learning.
- Optimizing agent - agent executing a solution improvement algorithms.

- Solution manager - agent responsible for managing the population of solutions.
- Attribute manager - agent responsible for the attribute selection coordination.

The parallel process of data reduction at sites is managed by the global level manager. Its role is to manage all stages of the learning process. As the first step the global manager identifies the distributed learning task that is to be coordinated and allocates optimizing agents to the local sites using the available agent migration procedure. Then the global manager initializes parallel execution of all subtasks, that is data reduction processes at local sites. When all the subtasks have been completed, solutions from the local levels are used to produce the global solution. Producing it requires that the global manager is equipped with skills needed to induce the global classifier. When the prototypes obtained from local sites are homogenous then the local prototypes are integrated and the global manager creates the global classifier (meta-classifier), using some machine learning algorithm. When the local level solutions are represented by heterogeneous set of prototypes then the global classifier can be induced by applying one of the meta-classifier strategies like, for example, bagging, AdaBoost, majority voting or some hybrid strategy.

Recently, Czarnowski and Jędrzejowicz [27] proposed to solve the distributed learning problem through replacing sampling by a data reduction process which can be carried-out at separate sites. The overall objective of data reduction is to process training data with a view to finding prototypes, which can replace the original training set during further steps of the supervised learning. Unfortunately, the above outlined process of learning from the distributed data can easily become more complicated. Main reasons for this are possible differences among the selected attributes of the distributed datasets introduced during the data reduction process carried out by simultaneously reducing datasets at each local site, since these are reduced in two dimensions, i.e. instance and attribute. The approach was an extension of the idea introduced in the paper of Czarnowski and Jędrzejowicz [26] where the instance selection is carried out independently at each site but the attribute selection is coordinated through the process of interaction and collaboration between agents based on a specialized strategy for agents collaborating and deciding on the winning set of attributes.

The idea is based on the assumption that prototypes are selected independently at each site through applying an agent-based population learning algorithm where datasets are reduced simultaneously in two dimensions. If the resulting global set of prototypes obtained through simple integration of the local site results is not homogenous it is suggested to apply the consensus method seeking to assure that a common set of attributes to be used at both - the local and the global levels is finally obtained.

The formal description of the problem of learning from the distributed data can be stated as follows: Given the distributed datasets $D_1, \ldots, D_K$, where examples are described by the sets of attributes $A_1, \ldots, A_K$ (where $\forall A_{i:i=1,\ldots,K} = \{a_1, \ldots, a_n\}$, a_i is an attribute and n is the number of attributes), a set of hypotheses H, a performance criterion F, the learning algorithm L outputs a hypothesis $h \in H$ that optimizes F. The task of the distributed learner L is to output the hypothesis $h \in H$ that optimizes performance criterion F (e.g. function of accuracy of classification, complexity of the hypothesis, classification cost or classification error) using data sets $D_1,\dots,D_K$ located in K sites. It should also be noted that when the distributed datasets are homogenous all attributes are presented at each site. i.e. $\exists_{i,j:i,j=1,\dots,K} A_i = A_j$, otherwise the datasets are called heterogeneous, i.e. $\exists_{i,j:i,j=1,\dots,K} A_i \neq A_j$. For the purpose of the consensus-based distributed learning it was assumed that the distributed datasets $D_1,\dots,D_K$ are homogenous.

The data reduction process aims at identifying and eliminating irrelevant and redundant information, and finding prototypes or regularities within certain attributes, allowing to induce the so-called prototypes or reference vectors. Main features of the data reduction processes, including data reduction for the distributed learning, are discussed in a detailed manner in [17]. The general definition of the data reduction problem can be formalized as follows: Given a learning algorithm L, and a dataset D with attributes described by an attribute set A, the optimal prototype dataset, S_{opt}, is a subset of the dataset D, where each example is described by a set of $A' \subset A$, such that the performance criterion of the learning algorithm L is maximized.

It can be easily observed that a good set of prototypes has the following properties:

- Firstly, the cardinality of the reduced dataset is smaller than the cardinality of the original, non-reduced dataset.
- Secondly, the reduced dataset assures maximum or acceptable classification quality with respect to the classifier induced using such a reduced data set. Classification quality is measured using a criterion or criteria provided by the user.

When the data reduction is considered with respect to physically distributed repositories then, at a final stage, datasets $D_1,\dots,D_K$ are replaced by the reduced datasets $S_1,\dots,S_K$ of local patterns such that $\forall_{i:i=1,\dots,K} S_i \subset D_i$. Thus the goal of data reduction is to find subset S_i from given D_i. In such case, the task of the distributed learner L is to output a hypothesis $h \in H$ optimizing F using datasets $S_1,\dots,S_K$, such that $\forall_{i:i=1,\dots,K} S_i \subset D_i$.

The process of learning from the distributed data is even more complex when the reduction is carried out in both dimensions. In such case data reduction can result in the reduced datasets, which are not necessarily homogenous i.e. $\forall_{i:i=1,\dots,K} A_i^{'} \subset A_i$ and it can be true that $\exists_{i,j:i,j=1,\dots,K} A_i^{'} \neq A_j^{'}$. In such case some techniques would be required to deal with the situation. One possible approach is applying a special combined strategy which is responsible for combining and integrating a number of classifiers learned from heterogeneous sets of prototypes available at the global level (see comparison of various strategies for combining classifiers in [23]). Another possible approach is to assure that prototypes obtained at each local site are homogenous in the discussed sense, i.e., characterized by identical set of attributes. In this paper it is suggested that applying the consensus method can lead to the desired attribute homogeneity. The consensus method is used to arrive at a common set of attributes for all the distributed datasets which best represents these datasets. Based on the consensus, it is possible to build one set of attributes out of many returned attribute sets from various distributed sites. The resulting homogenous attribute set is called a consensus of the attribute set. An overview of the consensus method and a discussion of different criteria for a consensus choice can be found in papers of Nguyen [56] and [57].

Within the above described setting, learning is carried-out in two stages, both involving cooperation between agents:

- Local stage, in which the selection of prototypes from the distributed data takes place. In this case the population-based approach with optimization procedures implemented as an asynchronous team of agents (A-Team), is used to select prototypes by instance selection and/or removing irrelevant attributes. Within the A-Team multiple agents achieve an implicit cooperation by sharing a population of solutions, also called individuals, to the problem to be solved. An A-Team can also be defined as a set of agents and a set of memories, forming a network in which every agent remains in a closed loop. All the agents can work asynchronously and in parallel. Agents cooperate to construct, find and improve solutions which are read from the shared, common memory. In our case the shared memory is used to store a population of solutions to the data reduction problem encountered at a given local site. Each solution is represented by a set of prototypes i.e. by the compact representations of the original datasets available at the considered site. A feasible solution to the data reduction problem at a local site is encoded as a string consisting of numbers of selected reference instances and numbers of selected attributes.
- Global stage, consisting of integrating or pooling the selected prototypes and producing the global learning model.

The approach allows also to deal with several local level data reduction problems solved in parallel. At the local level, that is, at the distributed data sources, agent-based population learning data reduction algorithms are executed in parallel. At the local level the proposed approach provides instance and attribute reduction capabilities integrated with the classifier learning process. It is expected that such an integration guarantees a high probability of getting the best set of prototypes for producing the global learning model at the global level. An important feature of the discussed solution is A-Teams ability to select instances and attributes in cooperation between agents thus assuring a homogenous set of prototypes at the global level. In this case, the instance selection is carried-out independently at each site through applying the agent-based population search but the attribute selection is managed and coordinated through the process of interaction and collaboration between agents.

The process of solving the data reduction problem is managed by the global manager, which is activated as the first one. The global manager is responsible for managing all stages of the distributed learning. Than the global manager runs in parallel all subtasks (i.e. data reduction processes) at local sites. When all the subtasks have been solved, solutions from the local level are used to obtain the global solution. Thus, the global manager creates the global set of prototypes by integrating local solutions and finally producing the global classifier, called also the meta-classifier.

The process of solving the data reduction is carried out by optimizing agents and the solution manager. Each optimizing agent is an implementation of a certain improvement algorithm, and the problem of data reduction at local sites is solved by A-Team, that is a team of optimizing agents possibly of different kinds supervised by the solution manager. The solution manager, is responsible for organizing the data reduction process at a local site through managing the population of solutions called individuals and updating them when appropriate. Each solution manager is also responsible for selecting the best obtained solution of the supervised local data reduction problem. During the data reduction process the solution manager continues reading individuals (solutions) from the common memory and storing them back after attempted improvement until a stopping criterion is met. During this process the solution manager keeps sending single individuals (solutions) from the common memory to optimizing agents. Solutions forwarded to optimizing agents for improvement are randomly drawn by the solution manager from the common memory. Each optimizing agent tries to improve quality of the received solution and afterwards sends it back to the solution manager, which, in turn, updates common memory by replacing a randomly selected individual with the improved one.

Each solution manager, after having supervised a predefined number of iterations within the data reduction process at a local site is obliged to send to the attribute manager a set of the candidate attributes. The attribute manager is responsible for coordination of the attribute selection while local data reduction problems are being solved in parallel. The attribute manager is responsible for finding the consensus solution, that is establishing the common set of attributes for all local data reduction problems. The attribute manager actions include receiving candidate attributes from solution mangers, collecting the candidate attributes from all solutions to local data reduction problems and finally deciding on the common set of attributes to be used at both - the local and the global levels. Thus, the attribute manager is an implementation of the consensus method, where based on answers from all solution managers the consensus set of attributes acceptable to a majority of solution managers is produced. Such an attribute set is called the winning set of attributes.

A procedure for determining the winning set of attributes is activated just after all local data reduction problems have been solved. Once the winning set of attributes have been chosen the attribute manager passes the outcome of its decision to all solution managers, whose role now is to update the respective individuals by correcting accordingly numbers of attributes in strings representing solutions in the current population. After this, the best solutions from each site are forwarded to the global level and merged into the global dataset by the global manager.

Another agent - the attribute manager is responsible for selecting the wining set of attributes. The process is again carried out in two stages. At the first stage candidate attribute sets from local sites are collected. The process is carried-out iteratively. At each iteration the candidate attribute sets are forwarded and collected by the attribute manager. Each candidate set received by the attribute manager is a list of the selected attributes representing the current best local solution and the fitness value of this solution. The fitness value is the estimated classification accuracy of the classifier induced from the reduced dataset calculated over the original training set. At the second stage a decision by consensus about the winning set of attribute is made.

The number of iterations at the first stage which is set by the user, determines the number of the candidate attribute sets received by the attribute manager. At the second stage selection of representatives amongst the received candidate sets is done for each local site independently. The selection is based on the fitness value of solutions from the received candidate sets. Candidate attribute sets are ordered from the best one to the worst forming a list. In case two or more candidate sets have the same fitness value, the value of the Hamming distance between each of them and the predecessor candidate attribute set on the list is calculated. The candidate set for which the above distance is greater is placed before the other one(s) at the ordered list of candidates. In case two or more of this way calculated Hamming distances are identical, only one randomly selected candidate stays on the list and the rest is removed. The algorithm for ordering candidate attribute sets is shown as Algorithm 1.

***Algorithm 1:** Evaluation and ordering of the candidate sets of attributes*

Input: $A_1^{'(1)},\ldots,A_i^{'(m)}$ - candidate attribute sets provided by i-th solution manager, m is the number of the received candidate sets; M - predefined number of iterations, that is the number of candidate sets from each local site to be considered at the second stage.
Output: X_i- the matrix of bits corresponding to M candidate attribute sets. The dimension of the matrix is $M \times n$, where n is the number of attributes in the original non-reduced set of attributes. The matrix denotes whether or not an attribute has been included in the respective candidate attribute set induced at the local level (1 - yes, 0- no).
1. Set $X_i := \emptyset$.
2. Create a list L of candidate sets of attributes $A_i^{'(j)}$ (where $j = 1,\ldots,m$) by ordering them according to values of their fitness. In case two or more candidate sets have the same fitness, the Hamming distance to the predecessor candidate attribute set on the list is calculated. The one with a greater distance is placed before the others. In case two or more of thus calculated Hamming distances are identical, only one randomly selected candidate stays on the list and the rest is removed.
3. For M first elements of L
4. Set rows of X_i according to attributes described by $A_i^{'(j)}$ (set an element of row to 1 if corresponding number of attribute is present in $A_i^{'(j)}$, otherwise set an element of row to 0).
5. End For.

At the second stage the winning set of attributes is selected through calculating the consensus solutions. According to Danilowicz and Nguyen [29] the consensus problem for solving inconsistency of the replicated data can be formulated as follows: given a set of versions of such data, one should select one version best representing the whole set. The selected version is called a consensus. In our case a consensus solution is produced from data stored in matrices $X_{i:i=1,\ldots,K}$. The idea of the proposed approach adopted from the paper of Sliwko and Nguyen [71] is to evaluate the weight of all candidate set of attributes on the basis of their frequency and positions within the respective matrices $X_{i:i=1,...,K}$. A consensus-based procedure for selection of the winning set of attributes is shown as Algorithm 2 given below. This algorithm guarantees that the winning set of attributes has the following properties:

- It is most similar to all other considered candidate attribute sets.
- It does not much differ from all other considered candidate sets.
- The differences between the consensus (the winning set of attributes) and all other considered candidate sets are as small as possible.

It should be noted that the winning set of attributes may contain only selected attributes from the consensus solution. In the proposed approach the number of attributes selected to the winning set of attributes depends on the acceptance level, which determines how many attributes from the original set of attributes should be accepted.

Algorithm 2: *A consensus-based procedure for selection of the winning set of attributes*

Input: $X_1,\ldots,X_K$ -matrices containing elements of candidate sets of attributes; where K is the number of distributed datasets and the number of solution managers; α - a value of the acceptance level set by the user.
Output: A'' - the winning set of attributes.
1. Set $A'' := \emptyset$.
2. Set $C := [a_1,\ldots,a_n]$, which is a vector representing the consensus solution, where n is the number of attributes.
3. Set $X := X_1 \cup X_2 \cup \ldots \cup X_K$, which has the size equal to $(M \cdot K) \times n$.
4. Set the vector $\bar{x} := [\bar{x}_1,\ldots,\bar{x}_n]$, where $\forall_{i:i=1,\ldots,n}\bar{x}_i = \frac{\sum_{k=1}^{M\cdot K} x_{ik}}{M\cdot K}$, $x_{ik} \in X$ and each x_i corresponds to a_i.
5. Sort elements in $\bar{x}$ in descending order. Sort elements in C accordingly to $\bar{x}$.
6. For all $x_i \in \bar{x}$ do
7. If $x_i \geq \alpha$ then add $a_i \in C$ to A'' (α should not be greater than the average value in $\bar{x}$).
8. End For.

The approach has been validated by means of computational experiment. The goal of the experiment was to evaluate how the proposed consensus method used to select a common set of attributes can influence the performance of the global classifier induced from the set of prototypes selected from the separate distributed sites. Classification accuracy of the global classifiers obtained using the above described approach to selecting the winning set of attributes have been compared with:

- Results obtained by pooling together all instances from the distributed databases, without data reduction, into the centralized database,
- Results obtained by pooling together all instances selected from distributed databases through the instance reduction procedure only,
- Results obtained by applying the static attribute selection procedure for distributed attribute selection proposed by Czarnowski and Jędrzejowicz [26].

Generalization accuracy has been used as the performance criterion. The learning tool used was C4.5 algorithm [63].The experiment involved four datasets - *customer*

(24000 instances, 36 attributes, 2 classes), *adult* (30162, 14, 2), *waveform* (30000, 21, 2) and *shuttle* (58000, 9, 7). For the first two datasets the reported classification accuracies are respectively 75.53% and 84.46% [30]. The above datasets have been obtained from UCI Machine Learning Repository [5] and EUNITE [30]. The reported computational experiment was based on the ten cross validation approach. At first, the available datasets have been randomly divided into the training and test sets in approximately 9/10 and 1/10 proportions. The second step involved the random partition of the previously generated training sets into the training subsets each representing a different dataset placed in the separate location. Next, each of the obtained datasets has been reduced. The reduced subsets have been then used to compute the global classifier using the proposed strategies. The above scheme was repeated ten times, using a different dataset partition as the test set for each trial.The original data set was randomly partitioned into, respectively, the 2, 3, 4 and 5 multi-databases of approximately similar size.

Table 1 Computational experiment results (average accuracy - in %, obtained by the C4.5 classifier)

	Number of the distributed data sources			
Problem	2	3	4	5
	Selection of reference instances at the local level only			
customer	68.45(+/-0.98)	70.4(+/-0.76)	74.67(+/-2.12)	75.21(+/-0.7)
adult	86.2(+/-0.67)	87.2(+/-0.45)	86.81(+/-0.51)	87.1(+/-0.32)
waveform	75.52(+/-0.72)	77.61(+/-0.87)	78.32(+/-0.45)	80.67(+/-0.7)
shuttle	99.95(+/-0.02)	99.92(+/-0.02)	99.98(+/-0.01)	99.96(+/-0.02)
average	82.53	83.78	84.95	85.74
	Static attributes election strategy			
customer	70.12(+/-1.28)	71.22(+/-1.46)	72.1(+/-1.1)	73.21(+/-0.7)
adult	86(+/-1.02)	85.3(+/-1.15)	87.1(+/-0.9)	87(+/-0.9)
waveform	76.12(+/-0.94)	78.2(+/-0.91)	78.32(+/-1)	80.37(+/-1.1)
shuttle	97.54(+/-1.2)	98.3(+/-1.1)	99.1(+/-1)	98.41(+/-1.1)
average	82.45	83.26	84.16	84.75
	Consensus-based attributes election			
customer	70.89(+/-0.61)	73.65(+/-0.84)	72.81(+/-1.02)	76.14(+/-0.62)
adult	87.04(+/-0.84)	86.5(+/-0.77)	88.62(+/-0.67)	88.04(+/-0.7)
waveform	74.58(+/-0.7)	80.02(+/-1.21)	81.51(+/-1.1)	80.54(+/-1.21)
shuttle	97.43(+/-0.41)	98.32(+/-0.97)	99.1(+/-0.8)	99.32(+/-0.72)
average	82.49	84.62	85.51	86.01

Source: [27].

The experiment results are shown in Table 1. These results have been averaged over ten cross validation runs. For the proposed approach the value of the acceptance level has been set to $\frac{\tilde{x}}{2}$, where $\tilde{x}$ is the average value in $\bar{x}$ (see Algorithm 2). The number of candidate sets to be stored for further consideration from each solution managers corresponding to the number of iterations in the procedure of ordering the candidate attribute sets, has been set to 10.

Generally, it should be noted that data reduction in two dimensions (selection of reference instances and attributes) assures better results in comparison to data reduction only in one dimension i.e. instance dimension. It has been also confirmed that learning classifiers from distributed data and performing data reduction at the local level produces reasonable to very good results in comparison with the case in which all instances from distributed datasets are pooled together.For example, pooling all instances from the distributed datasets assures classification accuracy of 73.32%(+/-1.42), 82.43%(+/-1.03), 71.01%(+/-0.8) and 99.9%(+/-0.03) for customer, adult, waveform and shuttle datasets respectively. On the other hand, the global classifier based on instance selection only assures classification accuracy of 75.21%, 87.1%, 80.67% and 99.96%. These results can still be considerably improved using the consensus-based approach to attribute selection, assuring classification accuracy of 76.14%, 88.04%, 88.54% and 99.32% respectively for the investigated datasets. These results also show that the consensus-based approach to attribute selection is competitive as compared with an earlier approach based on the static attribute selection strategy [26].

5 Conclusions

Main focus of this review is using agent technology in the field of machine learning with a particular interest on applying agent-based solutions to supervised learning. Some references are also made with respect to applying machine learning solutions to support agent learning. The review allows to formulate the following observations:

- Machine learning and agent technology are becoming more and more interelated bringing an important advantages to both fields.
- Machine learning can be seen as a prime supplier of learning capabilities for agent and multiagent systems.
- Agent technology have brought to machine learning several capabilities including parallel computation, scalability and interoperability.
- Agent-based solutions and techniques when applied to machine learning have proven to produce a synergetic effect originating from the collective intelligence of agents and a power of cooperative solutions generated through agent interactions.

Summary of the reviewed approaches to learning agents and machine learning with agents ideas and solutions is shown in Tables 2-3.

In Table 4 several applications of the A-Team technology to machine learning are summarized.

Future research should help to further integrate both fields - agent technology and machine learning. Agent based solutions could be used to develop more flexible and adaptive machine learning tools. Collective computational intelligence techniques can be used to effectively solve computationally hard optimization and decision problems inherent to many supervised learning techniques and data reduction

Table 2 Summary of the reviewed approaches to learning agents

Field	Scope	Example solutions/ideas
Learning agents	Techniques	Multiagent reinforcement learning - Busoniu et al. [13]
		Connection of the theory of automata with the multiagent reinforcement learning - Nowe et al. [58]
		Learning in multiagent systems and game theory - Shoham et al. [68]
		Data mining techniques supporting agents intelligence - Symeonidis et al. [75]
		Generic model for building adaptive agents using temporal difference methods - Preux et al. [61]
		Designing agents with machine learning capabilities - Sardinha et al. [67]
		MAS architecture based on connectionist learning - Rosaci [66]
		Stigmetry and entropy in learning automata based multiagent systems - Masoumi and Meybodi [53]
		Utility-based learning and strategic learning - Boylu et al. [11], Loizos [50]
		Probably Approximately Correct semantics to deal with missing information during the learning phase - Loizos [50]
		Multi-goal Q-learning algorithm for cooperative teams - Li et al. [48]
	Applications	Engineering applications and multiagent learning - Mannor and Shamma [52]
		Reinforcement learning for agent-based production scheduling - Wang and Usher [81]
		Case-based reinforcement learning algorithm for dynamic inventory control in a multi-agent supply-chain system - Jiang and Sheng [45]
		Support vector machine based multiagent ensemble learning for credit risk evaluation - Yu et al. [83]
		Agent Academy, an integrated development framework supporting design and control of multi-agent systems - Symeonidis et al. [76]

problems. Most promising direction for future research seems integration of machine learning and agent technology with a view to obtain effective solutions to the distributed learning problems. On the other hand more compact and reliable machine learning techniques are needed to equip agents with better learning capabilities.

Table 3 Summary of the reviewed approaches to machine learning with agents

Field	Scope	Example solutions/ideas
Machine learning	Techniques	On-line reinforcement learning system using a mixture of Bayesian networks - Kitakoshi et al. [46] Multiple heterogeneous, intelligent agents employing a different machine learning technique - Gifford [32] Neural network-based multi-agent classifier system - Quteishat et al. [64] Evolutionary Dynamics as a model for Q-learning in stochastic games - Hoenl and Tuyls [35] Collaborative machine learning framework - Hofmann and Basilico [34] Learning classifier systems (LCS) - Holland [36] Analysis of the properties of LCSs and extensions - [6], [7], [12] and [82] Mutual information LCS - Smith et al. [72] GAssist, a Pittsburgh-style Learning Classifier System - Bacardit and Krasnogor [8] Ensembles constructed from simple agents represented by expression trees induced using Gene Expression Programming - Jędrzejowicz and Jędrzejowicz [42] and [43] Ant Colony System-based framework for fuzzy data mining - Hong et al. [37], Parpinelli et al. [59], Cordon and Herrera [14] An agent paradigm as a tool for integration of different techniques into an effective strategy of learning from data - Zhang et al. [85]
	Applications	Agent Grid Intelligent Platform - Raicevic [65] Learning Intelligent Distribution Agent - Negatu et al. [55] Agent-based architectures for solving the distributed learning problems - Papyrus [62], MALE [69], ANIMALS [73], MALEF [79], EMADS [3] Personalized and intelligent information routing of online news - Fan et al. [31] Learning information agent for the identification of web documents - Pazzani and Billsus [60] Symbolic transducers and recurrent neural preference machines to support mining and classifying textual information - Arevian et al. [4] Detecting and responding to intrusions - Abraham et al. [1], Jansen [39], Moskovitch et al. [54] Case-based reinforcement learning algorithm (CRL) for dynamic inventory control in a multi-agent supply-chain system - Jiang and Sheng [45] Heterogeneous multiagent system for solving the pursuit problem - Ishiwaka [38] Multiagent machine learning and classifier combination to learn rock facies sequences - Gifford and Agah [33] Multiagent data mining approach for brain modeling - Zhang and Zhang [84] Agent technology integrated with data mining for e-learning - Acampora et al. [2] Multi-agent framework for data mining in electromyography - multi-agent framework for data mining in electromyography

Table 4 Summary of the reviewed approaches to machine learning with agents

ML area	Example solutions/ideas
Framework/environment	JADE-Based A-Team middleware environment (JABAT) - Barbucha et al. [10]
Supervised learning	Cascade correlation learning architecture - Czarnowski and Jędrzejowicz [19]
	Training the feed-forward artificial neural networks - Czarnowski and Jędrzejowicz [20]
Data reduction	Architectures, models and strategies for the A-Team based data reduction - Czarnowski and Jędrzejowicz [18], [21], [22], [16], [25]
	Cosensus-based distributed data reduction - Czarnowski and Jędrzejowicz [27]
Clustering	Similarity coefficient approach - Czarnowski and Jędrzejowicz [18], [21]
Distributed learning	Learning Classifiers from Distributed Data - Czarnowski and Jędrzejowicz [15], [17], [23], [26] and [28]
Instance selection	Selection through agents collaboration - Czarnowski and Jędrzejowicz [26]

References

1. Abraham, A., Jain, R., Thomas, J., Han, S.Y.: D-SCIDS: Distributed Soft Computing Intrusion Detection System. Journal of Network and Computer Applications 30, 81–98 (2007)
2. Acampora, G., Cadenas, J.M., Loia, V., Ballester, E.M.: A Multi-Agent Memetic System for Human-Based Knowledge Selection. IEEE Transactions on System, Men, and Cybernetics, Part A 41(5), 946–960 (2011)
3. Albashiri, K.A., Coenen, F., Leng, P.: EMADS: An Extendible Multi-agent DataMiner. Knowledge-Based Systems 22, 523–528 (2009)
4. Arevian, G., Wermter, S., Panchev, C.: Symbolic State Transducers and Recurrent Neural Preference Machines for Text Mining. International Journal of Approximate Reasoning 32, 237–258 (2003)
5. Asuncion, A., Newman, D.J.: UCI Machine Learning Repository. School of Information and Computer Science. University of California, Irvine (2007), `http://www.ics.uci.edu/learn/MLRepository.html`
6. Bacardit, J., Butz, M.V.: Data Mining in Learning Classifier Systems: Comparing XCS with GAssist. In: Kovacs, T., Llorà, X., Takadama, K., Lanzi, P.L., Stolzmann, W., Wilson, S.W. (eds.) IWLCS 2003. LNCS (LNAI), vol. 4399, pp. 282–290. Springer, Heidelberg (2007)
7. Bacardit, J., Garrell, J.M.: Bloat Control and Generalization Pressure Using the Minimum Description Length Principle for a Pittsburgh Approach Learning Classifier System. In: Kovacs, T., Llorà, X., Takadama, K., Lanzi, P.L., Stolzmann, W., Wilson, S.W. (eds.) IWLCS 2003. LNCS (LNAI), vol. 4399, pp. 59–79. Springer, Heidelberg (2007)

8. Bacardit, J., Krasnogor, N.: Empirical Evaluation of Ensemble Techniques for a Pittsburgh Learning Classifier System. In: Bacardit, J., Bernadó-Mansilla, E., Butz, M.V., Kovacs, T., Llorà, X., Takadama, K. (eds.) IWLCS 2006 and IWLCS 2007. LNCS (LNAI), vol. 4998, pp. 255–268. Springer, Heidelberg (2008)
9. Balter, J., Labarre-Vila, A., Zibelin, D., Garbay, C.: A Knowledge-driven Agent-centred Framework for Data Mining in EMG. C. R. Biologies 325, 375–382 (2002)
10. Barbucha, D., Czarnowski, I., Jędrzejowicz, P., Ratajczak-Ropel, E., Wierzbowska, I.: An Implementation of the JADE-base A-Team Environment. International Transactions on System Science and Applications 3(4), 319–328 (2008)
11. Boylu, F., Aytug, H., Koehler, G.J.: Principal-Agent Learning. Decision Support Systems 47, 75–81 (2009)
12. Bull, L., Kovacs, T.: Foundations of Learning Classifier Systems: An Introduction. Studies in Fuzziness and Soft Computing 183, 1–17 (2005)
13. Busoniu, L., Babuska, R., De Schutter, B.: A Comprehensive Survey of Multiagent Reinforcement Learning. IEEE Transactions on Systems, Man, and Cybernetics, Part C: Applications and Reviews 38(2), 156–172 (2008)
14. Cordon, J.C., Herrera, F.: Learning Fuzzy Rules Using Ant Colony Optimization. In: Proceedings of ANT 2000 International Workshop on Ant Algorithms, pp. 13–21 (2002)
15. Czarnowski, I.: Distributed Data Reduction through Agent Collaboration. In: Håkansson, A., Nguyen, N.T., Hartung, R.L., Howlett, R.J., Jain, L.C. (eds.) KES-AMSTA 2009. LNCS (LNAI), vol. 5559, pp. 724–733. Springer, Heidelberg (2009)
16. Czarnowski, I.: Prototype Selection Algorithms for Distributed Learning. Pattern Recognition 43, 2292–2300 (2010)
17. Czarnowski, I.: Distributed Learning with Data Reduction. In: Nguyen, N.T. (ed.) TCCI IV 2011. LNCS, vol. 6660, pp. 3–121. Springer, Heidelberg (2011)
18. Czarnowski, I., Jędrzejowicz, P.: An Approach to Instance Reduction in Supervised Learning. In: Coenen, F., Preece, A., Macintosh, A. (eds.) Research and Development in Intelligent Systems XX, pp. 267–282. Springer, London (2004)
19. Czarnowski, I., Jędrzejowicz, P.: An Agent-Based PLA for the Cascade Correlation Learning Architecture. In: Duch, W., Kacprzyk, J., Oja, E., Zadrożny, S. (eds.) ICANN 2005. LNCS, vol. 3697, pp. 197–202. Springer, Heidelberg (2005)
20. Czarnowski, I., Jędrzejowicz, P.: An Agent-based Approach to ANN Training. Knowlwedge-Based Systems 19, 304–308 (2006)
21. Czarnowski, I., Jędrzejowicz, P.: An Agent-Based Approach to the Multiple-Objective Selection of Reference Vectors. In: Perner, P. (ed.) MLDM 2007. LNCS (LNAI), vol. 4571, pp. 117–130. Springer, Heidelberg (2007)
22. Czarnowski, I., Jędrzejowicz, P.: An Agent-based Algorithm for Data Reduction. In: Bramer, M., Coenen, F., Petridis, M. (eds.) Research and Development of Intelligent Systems XXIV, pp. 351–356. Springer, London (2007)
23. Czarnowski, I., Jędrzejowicz, P.: A Comparison Study of Strategies for Combining Classifiers from Distributed Data Sources. In: Kolehmainen, M., Toivanen, P., Beliczynski, B. (eds.) ICANNGA 2009. LNCS, vol. 5495, pp. 609–618. Springer, Heidelberg (2009)
24. Czarnowski, I., Jędrzejowicz, P.: Cluster Integration for the Cluster-Based Instance Selection. In: Pan, J.-S., Chen, S.-M., Nguyen, N.T. (eds.) ICCCI 2010, Part I. LNCS (LNAI), vol. 6421, pp. 353–362. Springer, Heidelberg (2010)
25. Czarnowski, I., Jędrzejowicz, P.: An Approach to Data Reduction and Integrated Machine Classification. New Generation Computing 28, 21–40 (2010)
26. Czarnowski, I., Jędrzejowicz, P.: An agent-based framework for distributed Learning. Engineering Applications of Artificial Intelligence 24, 93–102 (2011)

27. Czarnowski, I., Jędrzejowicz, P.: A Consensus-Based Approach to the Distributed Learning. In: Proceedings IEEE SMC 2011, Anchorage, pp. 936–941 (2011)
28. Czarnowski, I., Jędrzejowicz, P., Wierzbowska, I.: An A-Team Approach to Learning Classifiers from Distributed Data Sources. International Journal of Intelligent Information and Database Systems 4(3), 245–263 (2010)
29. Danilowicz, C., Nguyen, N.T.: Consensus Methods for Solving Inconsistency of Replicated Datain Distributed Systems. Distributed and Parallel Databases 14, 353–369 (2003)
30. The European Network of Excellence on Intelligence Technologies for Smart Adaptive Systems. EUNITE World Competition in domain of Intelligent Technologies (2002), `http://neuron.tuke.sk/competition2`
31. Fan, W., Gordon, M., Pathak, P.: An integrated Two-stage Model for Intelligent Information Routing. Decision Support Systems 42(1), 362–374 (2006)
32. Gifford, C.M.: Collective Machine Learning: Team Learning and Classification in Multi-Agent Systems. Ph.D. dissertation, University of Kansas (2009)
33. Gifford, C.M., Agah, A.: Collaborative Multi-agent Rock Facies Classification from Wireline Well Log Data. Engineering Applications of Artificial Intelligence 23, 1158–1172 (2010)
34. Hofmann, T., Basilico, J.: Collaborative Machine Learning. In: Hemmje, M., Niederée, C., Risse, T. (eds.) From Integrated Publication and Information Systems to Information and Knowledge Environments. LNCS, vol. 3379, pp. 173–182. Springer, Heidelberg (2005)
35. Hoen, P.J., Tuyls, K.: Analyzing Multi-agent Reinforcement Learning Using Evolutionary Dynamics. In: Boulicaut, J.-F., Esposito, F., Giannotti, F., Pedreschi, D. (eds.) ECML 2004. LNCS (LNAI), vol. 3201, pp. 168–179. Springer, Heidelberg (2004)
36. Holland, J.H.: Escaping Brittleness: The possibilities of General-Purpose Learning Algorithms Applied to Parallel Rule-Based Systems. In: Michalski, R.S., Carbonell, J.G., Mitchell, T.M. (eds.) Machine Learning, An Artificial Intelligence Approach, vol. II, pp. 593–623. Morgan Kaufmann, Palo Alto (1986)
37. Hong, T.-P., Tung, Y.-F., Wang, S.-L., Wu, M.-T., Wu, Y.-L.: An ACS-based Framework for Fuzzy Data Mining. Expert Systems with Applications 36, 11844–11852 (2009)
38. Ishiwaka, Y., Sato, T., Kakazu, Y.: An Approach to the Pursuit Problem on a Heterogeneous Multiagent System Using Reinforcement Learning. Robotics and Autonomous Systems 43, 245–256 (2003)
39. Jansen, W.A.: Intrusion Detection with Mobile Agents. Computer Communications 25, 1392–1401 (2002)
40. Jędrzejowicz, P.: A-Teams and Their Applications. In: Nguyen, N.T., Kowalczyk, R., Chen, S.-M. (eds.) ICCCI 2009. LNCS, vol. 5796, pp. 36–50. Springer, Heidelberg (2009)
41. Jędrzejowicz, P.: Machine Learning and Agents. In: O'Shea, J., Nguyen, N.T., Crockett, K., Howlett, R.J., Jain, L.C. (eds.) KES-AMSTA 2011. LNCS (LNAI), vol. 6682, pp. 2–15. Springer, Heidelberg (2011)
42. Jędrzejowicz, J., Jędrzejowicz, P.: A Family of GEP-Induced Ensemble Classifiers. In: Nguyen, N.T., Kowalczyk, R., Chen, S.-M. (eds.) ICCCI 2009. LNCS (LNAI), vol. 5796, pp. 641–652. Springer, Heidelberg (2009)
43. Jędrzejowicz, J., Jędrzejowicz, P.: Two Ensemble Classifiers Constructed from GEP-Induced Expression Trees. In: Jędrzejowicz, P., Nguyen, N.T., Howlet, R.J., Jain, L.C. (eds.) KES-AMSTA 2010, Part II. LNCS (LNAI), vol. 6071, pp. 200–209. Springer, Heidelberg (2010)

44. Jennings, N., Sycara, K., Wooldridge, M.: A Roadmap of Agent Research and Development. Autonomous Agents and Multi-Agent Systems 1, 7–38 (1998)
45. Jiang, C., Sheng, Z.: Case-based Reinforcement Learning for Dynamic Inventory Control in a Multi-agent Supply-chain System. Expert Systems with Applications 36, 6520–6526 (2009)
46. Kitakoshi, D., Shioya, H., Nakano, R.: Empirical Analysis of an On-line Adaptive System Using a Mixture of Bayesian Networks. Information Science 180, 2856–2874 (2010)
47. Klusch, M., Lodi, S., Moro, G.: Agent-Based Distributed Data Mining: The KDEC Scheme. In: Klusch, M., Bergamaschi, S., Edwards, P., Petta, P. (eds.) Intelligent Information Agents. LNCS (LNAI), vol. 2586, pp. 104–122. Springer, Heidelberg (2003)
48. Li, J., Sheng, Z., Ng, K.C.: Multi-goal Q-learning of Cooperative Teams. Expert Systems with Applications 38, 1565–1574 (2011)
49. Liau, C.J.: Belief, Information Acquisition, and Trust in Multi-agent Systems - A modal Logic Formulation. Artificial Intelligence 149(1), 31–60 (2003)
50. Loizos, M.: Partial Observability and Learnability. Artificial Intelligence 174, 639–669 (2010)
51. Luo, J., Wang, M., Hu, J., Shi, Z.: Distributed Data Mining on Agent Grid: Issues, Platform and Development Toolkit. Future Generation Computer Systems 23, 61–68 (2007)
52. Mannor, S., Shamma, J.S.: Multi-agent Learning for Engineers. Artificial Intelligence 171, 417–422 (2007)
53. Masoumi, B., Meybodi, M.R.: Speeding up Learning Automata Based Multiagent Systems Using the Concepts of Stigmergy and Entropy. Expert Systems with Applications 38(7), 8105–8118 (2011)
54. Moskovitch, R., Elovici, Y., Rokach, L.: Detection of Unknown Computer Worms Based on Behavioral Classification of the Host. Computational Statistics and Data Analysis 52, 4544–4566 (2008)
55. Negatu, A., D'Mello, S., Franklin, S.: Cognitively Inspired Anticipatory Adaptation and Associated Learning Mechanisms for Autonomous Agents. In: Butz, M.V., Sigaud, O., Pezzulo, G., Baldassarre, G. (eds.) ABiALS 2006. LNCS (LNAI), vol. 4520, pp. 108–127. Springer, Heidelberg (2007)
56. Nguyen, N.T.: Using Distance Functions to Solve Representation Choice Problems. Fundamenta. Informaticae 48, 295–314 (2001)
57. Nguyen, N.T.: Consensus System for Solving Conflicts in Distributed Systems. Information Science 147, 91–122 (2002)
58. Nowé, A., Verbeeck, K., Peeters, M.: Learning Automata as a Basis for Multi Agent Reinforcement Learning. In: Tuyls, K., Hoen, P.J., Verbeeck, K., Sen, S. (eds.) LAMAS 2005. LNCS (LNAI), vol. 3898, pp. 71–85. Springer, Heidelberg (2006)
59. Parpinelli, R.S., Lopes, H.S., Freitas, A.A.: An Ant Colony Based System for Data Mining: Application to medical data. In: Proceedings of Genetic and Evolutionary Computation Conference, pp. 791–798 (2001)
60. Pazzani, M., Billsus, D.: Learning and Revising User Profiles: The Identification of Interesting Web Sites. Machine Learning 27(3), 313–331 (1997)
61. Preux, P., Delepoulle, S., Darcheville, J.-C.: A Generic Architecture for Adaptive Agents Based on Reinforcement Learning. Information Science 161, 37–55 (2004)
62. Prodromidis, A., Chan, P.K., Stolfos, S.J.: Meta-learning in Distributed Data Mining Systems: Issues and Approaches. In: Kargupta, H., Chan, P. (eds.) Advances in Distributed and Parallel Knowledge Discovery, vol. 3. AAAI/MIT Press, Menlo Park (2000)

63. Quinlan, J.R.: C4.5: Programs for Machine Learning. Morgan Kaufmann, SanMateo (1993)
64. Quteishat, A., Lim, C.P., Tweedale, J., Jain, L.C.: A Neural Network-based Multi-agent Classifier System. Neurocomputing 72, 1639–1647 (2009)
65. Raicevic, P.: Parallel Reinforcement Learning Using Multiple Reward Signals. Neurocomputing 69, 2171–2179 (2006)
66. Rosaci, D.: CILIOS: Connectionist Inductive Learning and Inter-ontology Similarities for Recommending Information Agents. Information Systems 32, 793–825 (2007)
67. Sardinha, J.A.R.P., Garcia, A., de Lucena, C.J.P., Milidiú, R.L.: A Systematic Approach for Including Machine Learning in Multi-agent Systems. In: Bresciani, P., Giorgini, P., Henderson-Sellers, B., Low, G., Winikoff, M. (eds.) AOIS 2004. LNCS (LNAI), vol. 3508, pp. 198–211. Springer, Heidelberg (2005)
68. Shoham, Y., Powers, R., Grenager, T.: If Multi-agent Learning is the Answer, what is the Question? Artificial Intelligence 171(7), 365–377 (2007)
69. Sian, S.: Extending Learning to Multiple Agents: Issues and a Model for Multi-Agent Machine Learning (Ma-Ml). In: Kodratoff, Y. (ed.) EWSL 1991. LNCS, vol. 482, pp. 440–456. Springer, Heidelberg (1991)
70. da Silva, J.C., Giannella, C., Bhargava, R., Kargupta, H., Klusch, M.: Distributed Data Mining and Agents. Engineering Applications of Artificial Intelligence 18, 791–807 (2005)
71. Sliwko, L., Nguyen, N.T.: Using Multi-agent Systems and Consensus Methods for Information Retrival in Internet. International Journal Itelligence Information and Databases Systems 1(2), 181–198 (2007)
72. Smith, R.E., Jiang, M.K., Bacardit, J., Stout, M., Krasnogor, N., Hirst, J.D.: A Learning Classifier System with Mutual-information-based Fitness. Evolutionary Intelligence 1(3), 31–50 (2010)
73. Stolfo, S., Prodromidis, L., Tselepis, S., Lee, W., Fan, D.W.: JAM: Java Agentsfor Metalearning over Distributed Databases. In: 3rd International Conference on Knowledge Discovery and Data Mining, pp. 74–81. AAAI Press, NewportBeach (1997)
74. Sutton, R.S., Barto, A.G.: Reinforcement Learning. An Introduction. MIT Press, Cambridge (1998)
75. Symeonidis, A.L., Chatzidimitriou, K.C., Athanasiadis, I.N., Mitkas, P.A.: Data Mining for Agent Reasoning: A synergy for Training Intelligent Agents. Engineering Applications of Artificial Intelligence 20, 1097–1111 (2007)
76. Symeonidis, A.L., Athanasiadis, I.N., Mitkas, P.A.: A Retraining Methodology for Enhancing Agent Intelligence. Knowledge-Based Systems 20, 388–396 (2008)
77. Takadama, K., Inoue, H., Shimohara, K., Okada, M., Katai, O.: Agent Architecture Based on an Interactive Self-reflection Classifier System. Artificial Life and Robotics 5, 103–108 (2001)
78. Talukdar, S., Baerentzen, L., Gove, A., De Souza, P.: Asynchronous Teams: Cooperation Schemes for Autonomous Agents. Journal of Heuristics 4(4), 295–321 (1998)
79. Tozicka, J., Rovatsos, M., Pechoucek, M., Urban, U.: MALEF: Framework for Distributed Machine Learning and Data Mining. International Journal of Intelligent Information and Database Systems 2(1), 6–24 (2008)
80. Tweedale, J., Ichalkaranje, N., Sioutis, C., Jarvis, B., Consoli, A., Phillips-Wren, G.E.: Innovations in Multi-agent Systems. Journal of Network and Computer Applications 30(3), 1089–1115 (2007)

81. Wang, Y.-C., Usher, J.M.: Application of Reinforcement Learning for Agent-based Production Scheduling. Engineering Applications of Artificial Intelligence 18, 73–82 (2005)
82. Wilson, S.W.: State of XCS Classifier System Research. In: Lanzi, P.L., Stolzmann, W., Wilson, S.W. (eds.) IWLCS 1999. LNCS (LNAI), vol. 1813, pp. 63–81. Springer, Heidelberg (2000)
83. Yu, L., Yue, W., Wang, S., Lai, K.K.: Support Vector Machine Based Multiagent Ensemble Learning for Credit Risk Evaluation. Expert Systems with Applications 37, 1351–1360 (2010)
84. Zhang, W.-R., Zhang, L.: A Multiagent Data Warehousing (MADWH) and Multiagent Data Mining (MADM) Approach to Brain Modeling and Neurofuzzy Control. Information Science 167, 109–127 (2004)
85. Zhang, S., Wu, X., Zhang, C.: Multi-Database Mining. IEEE Computational Intelligence Bulletin 2(1), 5–13 (2003)
86. Zhu, X., Bin, L., Wu, X., He, D., Zhang, C.: CLAP: Collaborative Pattern Mining for Distributed Information Systems. Decision Support Systems 52, 40–51 (2011)

Ant Colony Optimization for the Multi-criteria Vehicle Navigation Problem

Mariusz Boryczka* and Wojciech Bura

Abstract. The chapter describes the multi-agent ant-based vehicle navigation algorithms, which find multi-criteria optimal route between two points on the map. Presented are various versions of the algorithm, sequential and parallel, including GPU one. Various experiments performed on data of different size show the ability of presented algorithms to find good (near optimal) solutions for large real map. In turn, the parallel AVN algorithm is able to produce even better results in shorter time. Finally, we show that the presented approach may be adapted to run on GPUs and the algorithm's performance scales very well with growing number of multiprocessors.

1 Introduction

In recent years we witness the rapid growth in popularity and development of vehicle navigation systems. This is primarily due to the increasing availability of portable electronic devices equipped with a GPS module, which may be used for this purpose. One of the problems which is usually faced when dealing with navigation system is finding the optimal route between two selected points on the map. It does not mean that this route should be the shortest one. It should be optimal or quasi-optimal in term of user preferences.

The problem is known from the literature as Multi-criteria Shortest Path Problem, and it is proven to be NP-complete [13]. For multi-criteria combinatorial problems a single solution will very seldom be able to minimize (or maximize) all criteria, but rather there will be a set of compromise solutions. These solutions are called efficient non-dominated ones and are also referred to as Pareto optimal set.

Mariusz Boryczka · Wojciech Bura
Institute of Computer Science, University of Silesia, Będzińska 39,
43-100 Sosnowiec, Poland
e-mail: mariusz.boryczka@us.edu.pl, wojciech.bura@asseco.pl

* Corresponding author.

I. Czarnowski et al. (Eds.): Agent-Based Optimization, SCI 456, pp. 29–53.
DOI: 10.1007/978-3-642-34097-0_2

This chapter describes the ant-based vehicle navigation algorithm. The algorithm finds the optimal (or nearly optimal) route between two points on the map, taking into consideration user preferences for distance, traffic load, road width, risk of collision, quality and number of intersections. Parameterization of the algorithm is performed by setting the corresponding coefficients for the different criteria of optimality. Time of departure is also taken into account due to the fact that the weight of the individual segment of the route may have assigned different values at different times of day or night. It is important that the algorithm is able to propose a set of good solutions. We show that the ant-based approach can manage to solve such a problem.

Ant algorithms, in general, take inspiration from the behavior of real ant colonies to solve combinatorial optimization problems. They are based on a colony of artificial ants (agents) that work cooperatively and communicate through artificial pheromone trails. The artificial ant in turn is a simple, computational agent that tries to build feasible solutions to the problem tackled exploiting the available pheromone trails and heuristic information.

From the beginning of the researches on the ant algorithms scientists tried to parallelize them to achieve better results in quality and computation time. Most parallel implementations of ant algorithms are just parallelization of their standard form [2, 3, 8, 23, 28]. They differ only in whether the computations for the new pheromone matrix are done locally in all colonies or centrally by a master processor. Some authors also consider multi-colony algorithms [3, 16, 17, 18]. Different approaches for parallel ant algorithms are also described in [14]. Unfortunately, the obtained speedup values for different problems (especially TSP) and using 8-15 processors ranges from value less then 1 (which means parallel execution is worse than serial execution) up to about 3. Recently the promising results gave the application of the CUDA approach, where the ant algorithm was specifically adjusted to this architecture, and speedup of running time of algorithm for TSP implemented on GPU was about 20 [30]. This inspired us to use this idea for the parallel version of AVN. This new algorithm produced better results than its sequential predecessor, and high performance of this algorithm is very important because of its potential use in the on-line mode. It is significant that the latest mobile devices are equipped with a GPU capable of general calculations (for example PowerVR SGX Series5XT in iPhone and iPad). Therefore, the parallel version of the proposed algorithm also exploits OpenCL language, so it is able to leverage modern CUDA architecture.

Later in the chapter the experiments with the real data are presented. The data were collected from the OpenStreetMap system [20]. The results of the experiments show that the sequential ant based navigation algorithm is able to find good (near optimal) solutions for a large real map. It was also interesting, that the solution produced by the algorithm was regarded by many people as the best route. Experiments with various parallel versions of the ant colony vehicle navigation algorithm indicate its good susceptibility for parallelization.

The work is organized as follows. Section 2 shortly describes ant systems. Sections 3 and 4 present the original, sequential AVN algorithm and its improved version (NAVN), more efficient and capable of use for the real-word data. Section 5 describes various parallel processing approaches to ant colony optimization known from literature. Section 6 presents the basic parallel version of NAVN — PAVN, which runs in shared memory model. In section 7 the OpenCL version of PAVN algorithm is described (OCLAVN) and some algorithmic and programming issues that had to be solved during implementation are presented. Section 8 presents results of experiments with various versions of the AVN algorithm. Most of the conducted experiments where done using real data. Section 9 summarizes the work.

2 Ant Algorithms

Ant algorithms take inspiration from the behavior of real ant colonies to solve combinatorial optimization problems. They are based on a colony of artificial ants, that is, simple computational agents that work cooperatively and communicate through artificial pheromone trails [7]. The artificial ant is a simple, computational agent that tries to build feasible solutions to the problem tackled exploiting the available pheromone trails and heuristic information. It has some characteristic properties. It searches minimum cost feasible solutions for the problem being solved. It has a memory storing information about the path followed until that moment. It starts in the initial state and moves towards feasible states, building its associated solution incrementally. The movement is made by applying a transition rule, which is a function of the locally available pheromone trails and heuristic values, the ant's private memory and the problem constraints. During the construction procedure, an ant moves, it can update the pheromone trail associated to the edge. The construction procedure ends when any termination condition is satisfied, usually when an objective state is reached. Once the solution has been built, the ant can retrace the traveled path and update the pheromone trails on the visited edges.

In the ant colony optimization system (ACO) a virtual ant, being in a particular moment of time and at a certain stage of building the solution to the problem, selects the next step based on a specific transition rule. For this purpose it generates a random number q, $0 \leq q \leq 1$. If $q \leq q_0$ (q_0 — given algorithm's parameter), "the best" available decision is taken (deterministic — exploitation), otherwise the decision is taken at random (exploration), taking into account the probabilities calculated in accordance to formula (1) [11].

$$j = \begin{cases} \arg\max\limits_{r \in J_i^k} \{[\tau_{i,r}(t)] \cdot [\eta_{i,r}]^{\beta}\}, & \text{if } q \leq q_0 \text{ (exploitation)} \\ S, & \text{otherwise (exploration)}, \end{cases} \tag{1}$$

where:
$\tau_{i,j}$ – value of the prize, the degree of usefulness of the decision,
$\eta_{i,r}$ – heuristically estimated value of the quality of the transition from state i to state j, (for example, for the TSP problem it is the visibility of the city j from the city i)
β – importance level of $\eta_{i,r}$,
S – the next step (decision) drawn with probabilities:

$$p_{i,j}^k(t) = \begin{cases} \frac{\tau_{i,j}(t)\cdot[\eta_{i,j}]^\beta}{\sum\limits_{r\in J_i^k} \tau_{i,r}(t)\cdot[\eta_{i,r}]^\beta}, & \text{if } j \in J_i^k \\ 0, & \text{otherwise,} \end{cases}$$

where J_i^k is a set of decisions, which ant k can decide being in state i.

After each step, a virtual ant updates pheromone trail locally on the edge of their choice (the on-line step-by-step pheromone trail update). This procedure is also associated with its partial evaporation (instead of this, the pheromone evaporation may be performed after each cycle). It is done according to the formula (2).

$$\tau_{i,j}(t+1) = (1-\rho)\cdot\tau_{i,j}(t) + \rho\cdot\tau_0 \tag{2}$$

where:
ρ — pheromone evaporation coefficient, $0 \le \rho \le 1$,
τ_0 — initial pheromone trail value.

At the same time, the currently selected node (which leads to the chosen edge) is added to the TABU list (ant memory), which contains a list of nodes already visited by an ant in a given cycle.

After completing the full cycle the pheromone trail is updated globally on the edges which belong to the best solution found and pheromone level is changed (the on-line delayed pheromone trail update) using the following formula (3).

$$\tau_{i,j}(t+n) = (1-\gamma)\cdot\tau_{i,j}(t) + \gamma\cdot\frac{1}{L^+} \tag{3}$$

where transition decisions from i to j belong to the best solution, γ is the pheromone evaporation rate (if the evaporation is present here), and L^+ is length of the route.

Procedure ACOProc presents the structure of a generic ACO algorithm [9, 10].

The first step involves the initialization of the parameter values of the algorithm (e.g. the initial pheromone trail value associated to each transition of A, the number of ants in the colony, the weights in the probabilistic transition rule). The main procedure of the ACO manages the operation of the artificial ants, and the pheromone evaporation. It takes into consideration an ant memory L (TABU list), a set of constraints R defined for the problem and the parameters of the algorithm.

Procedure. ACOProc

```
begin
    Parameter-initialization;
    while termination-criterion-not-satisfied do
        foreach ant do                                    /* maybe in parallel */
            Initialize-ant;
            L = Update-ant-memory;
            while current-state ≠ target-state do
                P = Compute-transition-probabilities( A, L );
                next-state = Apply-ant-decision-policy( R );
                Move-to-next-state(next-state);
                if on-line-step-by-step-pheromone-update then
                    Deposit-pheromone-on-the-visited-edge;
                L = Update-internal-state;
            if on-line-delayed-pheromone-update then
                foreach visited-edge do
                    Deposit-pheromone-on-the-visited-edge;
            Release-ant-resources;
        Pheromone-evaporation;
```

3 AVN Algorithm and Its Modifications

There were proposed several systems for optimum route selection for car navigation systems, e.g. genetic algorithms [1, 15] or fuzzy logic [21, 29]. Another one is AVN — an ant-based algorithm approach to vehicle navigation [25, 26, 27]. This algorithm is based on the ant algorithm [7, 11]. It finds the optimal (or nearly optimal) routes between two points on the map, using user preferences for distance, traffic load, road width, risk of collision, quality and number of intersections. During calculations, also the time of departure is taken into account due to the fact that the weight of the individual segment of the route may have assigned different values at different time of day or night. It is important that we search for an algorithm which is able to propose not only the best (optimal) route, but a set of good solutions. Thus, it is possible to choose the alternative route when the unforeseeable situation (traffic jam, temporarily closed road etc.) occurs.

3.1 Problem Description

Finding the optimal path between two points on a map is a multicriterial problem of finding the shortest path in the graph, known in literature as an MSSP (Multiobjective Shortest Path Problem). Formally, this problem can be formulated as follows [22]:

Given a directed graph $G = (V,E)$, where V is set of vertices (nodes) and E the set of edges (arcs) with cardinality $|V| = n$ and $|E| = m$ and a d-dimensional function vector $c : E \rightarrow [\mathbb{R}^+]^d$. Each edge $e \in E$ is associated with a cost vector $c(e)$. A source vertex s and a sink vertex t are identified. A path p is a sequence of vertices and arcs from s to t. The cost vector $C(p)$ for linear functions of path p is the sum of the cost vectors of its edges, that is $C(p) = \sum_{e \in p} c(e)$. Given the two vertices s and t, let $P(s,t)$ denote the set of all paths from s to t in G. If all objectives are to be minimized, a path $p \in P(s,t)$ dominates a path $q \in P(s,t)$ if $\forall_{i \in d} C_i(p) \leq C_i(q)$ and we write $p \preceq q$. A path p is Pareto-optimal if it is not dominated by any other path and the set of nondominated solutions (paths) is called the Pareto-optimal set. The objective of the MSPP is to compute the set of nondominated solutions that is the Paretooptimal set $\mathbb{P} \in P(s,t)$ with respect to c.

This problem belongs to the class of NP-complete problems [12, 13] which encourages to use a heuristic methods that work quickly, but do not guarantee that the optimal solution is found. The Ant Colony Optimization algorithms are an example of such a methods.

3.2 *Original AVN Algorithm*

The Ant-based Vehicle Navigation algorithm (AVN) proposed by H.Salehinejad et al. [25, 26, 27] finds optimal route, which best fits preferences desired by the user. The set of user parameters consists of coefficients controlling the importance level of distance, width, number of intersections, traffic, risk and quality of the proposed route. It is based on the ant system introduced by Dorigo, Maniezzo and Colorni [6]. Steps of the AVN algorithm are as shows procedure AVNProc. Below we shortly describe the elements of mentioned algorithm.

First, initial setup of algorithm parameters is performed. Then ants are located at the starting point. Each ant is active until it reaches the destination point and is not blocked at the intersection. An ant is blocked when there is no way to choose to continue the travel. In the next step, the probability of each possible next direct route is calculated based on the cost function calculated for each active ant. The probability of the move from node i to node j for ant k is calculated as:

$$p_{ij}^k = \begin{cases} \dfrac{\tau_{ij}^{\alpha} \prod_{l \in parameters} \xi_{ij_l}^{-2\alpha_l}}{\sum_{h \notin tabu_k} \tau_{ih}^{\alpha} \prod_{l \in parameters} \xi_{ih_l}^{-2\alpha_l}} & \text{if } j \in \text{tabu}_k \\ 0 & \text{otherwise} \end{cases} \tag{4}$$

where:

τ_{ij} — value of pheromone trail on edge from i to j,
α — coefficient that controls importance level of τ_{ij},
tabu_k — a set of unavailable nodes, already visited by the ant k,
ξ_{ij_l} — value of cost functions for parameter l for edge ij,
α_l — coefficient controlling the importance level of parameter l,

Procedure. AVNProc

```
begin
    Initialize;
    foreach loop do
        Locate Ants;
        foreach iteration do
            foreach ant do
                if ant is active then
                    Construct Probability;
                    Select Route;
                    Update TABU List;
                    Kill blocked ant;
        Value Ants;
        Award Winner Ants;
        Punish Loser Ants;
        Evaporate Pheromone;
    Select Best Optimized Direction;
```

and cost functions for all parameters for edge ij are: $\xi_{ij_{distance}}$, $\xi_{ij_{width}}$, $\xi_{ij_{traffic}}$, $\xi_{ij_{risk}}$ and $\xi_{ij_{quality}}$.

Based on the calculated probability ant k selects a route to go. To do this, a random value q in range $\langle 0.1\rangle$ is compared with parameter $Q \in \langle 0.1\rangle$, to choose an exploitation or an exploration:

$$j = \begin{cases} \arg\max_{h \in J_i^k}\{p_{ih}^k\} & \text{if } q \leq Q \text{ (exploitation)} \\ S & \text{otherwise (exploration)} \end{cases} \tag{5}$$

In next steps the node selected by the ant is added to its TABU list and if ant k arrives the destination or is blocked at the certain node, it is deleted from the active ant list. For each ant that reached the destination the complete cost ψ of the whole route is calculated and for all calculated costs ψ the average cost χ is worked out. If $\psi_k < \chi$ then ant is added to AWA (Award Winner Ants) list, otherwise it is added to PLA (Punish Loser Ants) list. In addition, for an ant (if any) with cost ψ lower then the cost of the global best solution then new best solution is remembered.

At the end of every loop updating of pheromone trail takes place. The pheromone trail on edge ij for ant k is updated as follows:

$$\tau_{ij}(t) = \begin{cases} \tau_{ij}(t-1) + \frac{av}{\psi_{ij}} & \text{if } k \in \text{AWA} \\ \tau_{ij}(t-1) \cdot pv & \text{if } k \in \text{PLA} \end{cases} \tag{6}$$

where av is the awarding coefficient, $av > 1$, and pv is the punishment coefficient, $0 < pv < 1$.

Then the global evaporation of the pheromone trail is performed as:

$$\tau_{ij}(t) = \rho \cdot \tau_{ij}(t-1) \tag{7}$$

where ρ — evaporation coefficient, $0 < \rho < 1$. After the assumed number of repetitions of the algorithm the global best solution is returned as the result.

3.3 *Improved AVN Algorithm*

On the basis of the AVN algorithm we prepared its slightly improved version (CAVN). The main change was a normalization procedure of each element of the cost function and the way of parameterization of user preferences. In the original algorithm particular α coefficients work the same way as user preferences and they play a role of the normalization. Thus, proper set-up of these parameters is not an easy task, and what is even more important, it is not comfortable for the user.

In the CAVN algorithm, the normalization coefficients are initially calculated at the beginning and they are then used during calculations of the cost function and probabilities. These values depend on the map only, so that they may be prepared before algorithm starts. Procedure CAVNProc presents steps of this algorithm. Below we describe new and modified elements of the CAVN algorithm (underlined in pseudo-code).

Procedure. CAVNProc

```
begin
    Prepare Normalization;
    Initialize;
    foreach loop do
        Locate Ants;
        foreach iteration do
            foreach ant do
                if ant is active then
                    Construct Probability;
                    Select Route;
                    Update TABU List;
        Value Ants;
        Award Winner Ants;
        Punish Loser Ants;
        Evaporate Pheromone;
        Update Q;
    Select Best Optimized Direction;
```

Prepare Normalization. In this step for each element of the cost function (distance, width, traffic, risk, quality and intersections) normalization coefficient η_l is calculated. Therefore user preference parameters have values in range $\langle 0,1\rangle$ and there is no need to adjust them to particular data on the map.

For calculations of the cost function maximum and minimum values of the user preferences are assigned to variables: $\max_{distance}$, $\max_{width}$, $\max_{traffic}$, $\max_{risk}$, $\max_{quality}$, and

$$\max_{all} = \max\{\max_{distance}, \max_{width}, \max_{traffic}, \max_{risk}, \max_{quality}\} \tag{8}$$

calculated. Then particular coefficients are calculated as follows:

$$\eta_{distance} = \frac{\max_{all}}{\max_{distance}} \tag{9}$$

$$\eta_{width} = \frac{\max_{all}}{\max_{width} / \min_{width}} \tag{10}$$

$$\eta_{traffic} = \frac{\max_{all}}{\max_{traffic}} \tag{11}$$

$$\eta_{risk} = \frac{\max_{all}}{\max_{risk}} \tag{12}$$

$$\eta_{quality} = \frac{\max_{all}}{\max_{quality} / \min_{quality}} \tag{13}$$

$$\eta_{intersection} = \max_{all} \tag{14}$$

Construct Probability. Here, normalization coefficients η_l are calculated (they are exploited in calculations of probabilities):

$$\xi_{ij_{distance}} = d(i,j) \cdot \eta_{distance} \tag{15}$$

$$\xi_{ij_{width}} = \frac{\max_{width}}{w(i,j)} \cdot \eta_{width} \tag{16}$$

$$\xi_{ij_{traffic}} = tf(i,j,t) \cdot \eta_{traffic} \tag{17}$$

$$\xi_{ij_{risk}} = r(i,j,t) \cdot \eta_{risk} \tag{18}$$

$$\xi_{ij_{quality}} = \frac{\max_{quality}}{q(i,j)} \cdot \eta_{quality} \tag{19}$$

Value Ants. Normalization coefficients are finally used in the complete cost calculation:

$$\xi^k_{intersection} = \mathrm{sizeof}(tabu_k) \cdot \eta_{intersection} \tag{20}$$

Award Winner Ants. In this step usage of parameter av was slightly modified:

$$\tau_{ij}(t) = \begin{cases} \tau_{ij}(t-1) \cdot av \text{ if } k \in \text{AWA} \\ \tau_{ij}(t-1) \cdot pv \text{ if } k \in \text{PLA} \end{cases} \tag{21}$$

where av is the awarding coefficient, $av > 1$, and pv is the punishment coefficient, $0 < pv < 1$.

Update Q. In this step parameter Q is decreased by coefficient φ ($\varphi \in \langle 0, 1 \rangle$) which is an algorithm's parameter:

$$Q(new) = Q(old) \cdot \varphi \tag{22}$$

This way in each loop we gradually reduce exploration on behalf of exploitation.

4 New Version of AVN — The Backtracking Algorithm

Running original (AVN) and improved (CAVN) algorithms on bigger datasets sometimes they did not bring any solution. The removal of blocked ants has proven to be too rigorous, and none of ants has reached the destination. As a result of our research a new ant algorithm (NAVN) has been constructed [4]. This algorithm is more similar to classical form of ACO algorithms. Results of computational experiments show that this algorithm is much more efficient and capable of use for the real-word data.

In the NAVN algorithm, killing the blocked ants is replaced by their returns from dead ends. Pheromone trail is updated both locally (when an ant is traveling on the map) and globally (after every loop) on the paths from the best solution. Additionally, while moving back a blocked ant highly reduces pheromone trial on the edge which led it to a node without a way out. This considerably reduces the probability of selecting this edge by other ants. Procedure NAVNProc presents the algorithm and below we describe its main steps:

Initialize. Setting the parameters of the algorithm. The initial amount of pheromone on edges is set to value τ_0.

Construct Probability. Calculation of the components of the probabilities is more similar to the classical ant algorithm and is expressed by the formula:

$$p_{ij}^{k} = \begin{cases} \dfrac{\tau_{ij} \cdot \left(\frac{1}{\psi_{ij}}\right)^{\beta}}{\sum_{h \notin \text{tabu}_k} \tau_{ih} \cdot \left(\frac{1}{\psi_{ih}}\right)^{\beta}} & \text{if} j \in \text{tabu}_k \\ 0 & \text{otherwise} \end{cases} \tag{23}$$

where: τ_{ij} — value of pheromone trail on the edge from i to j, β — coefficient controlling importance of the cost, ψ_{ij} — the cost of the edge from i to j calculated as follows:

$$\psi_{ij} = \sum_{l \in parameters} \xi_{ij_l} \cdot \alpha_l \tag{24}$$

Procedure. NAVNProc

```
begin
    Prepare Normalization;
    Initialize;
    foreach loop do
        Locate Ants;
        foreach iteration do
            foreach ant do
                if ant is active then
                    Construct Probability;
                    if ant has no move then Move Back;
                    else
                        Select Route;
                        Update TABU List;
        Value Ants;
        Award Best Solution;
        Punish Loser Ants;
        Modify Q;
    Select Best Optimized Direction;
```

where ξ_{ij_l} is the value of the cost function for parameter l and edge (i, j), calculated as in the CAVN algorithm, and α_l is the coefficient controlling importance of parameter l and is assigned by the user ($0 < \alpha_l < 1$).

Move Back. If ant k is locked (there is no edge it can travel), it makes one step back. The edge used to go back is added to the list called *blindEdgesk*$_k$, which contains the list of forbidden edges which will be no more chosen by ant k in current loop. The pheromone trail on the edge is updated as follows:

$$\tau_{ij}(new) = \tau_{ij}(old) \cdot bv \tag{25}$$

where bv is the blind edge pheromone change parameter.

Select Route. Choosing an edge is performed the same way as in previous algorithms but after selecting the edge an ant updates a pheromone trail according to the formula:

$$\tau_{ij}(new) = (1 - \rho) \cdot \tau_{ij}(old) + \rho \cdot \tau_0 \tag{26}$$

where: ρ is the trail evaporation coefficient ($0 \leq \rho \leq 1$).

Value Ants. Solutions found by the ants are evaluated. If there is a solution better than the best one already found, it is saved as the new best solution. As in the previous algorithms, total cost includes the cost connected with the number of intersections:

$$\xi^k_{intersection} = \text{sizeof}(\text{tabu}_k) \cdot \eta_{intersection} \tag{27}$$

AwardBestSolution. The best solution is rewarded through the global pheromone trail updating rule according to the principle expressed by the formula:

$$\tau_{ij}(new) = (1-\rho)\cdot\tau_{ij}(old) + \gamma\cdot\rho\cdot\frac{\theta_{best}}{\psi_{best}} \tag{28}$$

where: θ_{best} — number of edges belonging to the best solution's path, ψ_{best} — the cost of the best solution, and γ — the parameter reinforcing the award for the best solution.

PunishNonActiveAnts. In this step, ants which failed to find a solution within a assumed number of iterations are punished — the pheromone trail on the edges belonging to their routes is decreased with the punishment coefficient $pv \in (0,1)$:

$$\tau_{ij}(new) = pv\cdot\tau_{ij}(old) \tag{29}$$

5 The Parallel ACO Approaches

Intuition suggests that multi-agent systems have great potential for parallelization. In the literature there are many well-known proposals of parallel versions of ACO algorithms. It is possible to specify certain characteristics common to most approaches [14]:

- Building solutions (including an evaluation of its quality) by an ant does not run on multiple processors — it is usually done on a single processor — because:
  - The process of design solutions usually proceeds sequentially, and it can not be divided into sections performed independently (assessment of quality of solutions can be time consuming and this process can be run in parallel).
  - This process is the smallest "executive" element of ACO, which must be executed sequentially (except in the case of hardware oriented ACO).
  - Sometimes it puts a lot of ants on a single processor treating them as a colony of cooperating ants.
- When assessing the quality of solutions, they are compared with the solutions obtained by the standard versions of the ACO, which are run independently on different processors, perhaps with different parameter values.

We can distinguish basic types of parallel ACO version [14]:

- Parallelization of standard version of ACO, or create a new version, taking into account the parallelism (e.g. exchanging information between processors less than after each iteration, which by the way causes the colonies to search different regions of the solution space).

- Centralized or decentralized approach to a pheromone matrix:
  - With a centralized approach — selected processor collects information about the solution or pheromone from other processors, updates and distributes pheromone tables to other processors (master-slave principle).
  - In a distributed approach — each processor has its own pheromone matrix, which is updated based on information obtained from other processors.
  - Synchronous or asynchronous communication between processors.
  - The use of uniform or heterogeneous processing units (distributed processing on multiple computers) — usual multi colony approach.
  - Hardware oriented ACO approaches (Figure 1 and Figure 2):
    · R-Mesh-ACO,
    · FPGA-ACO.

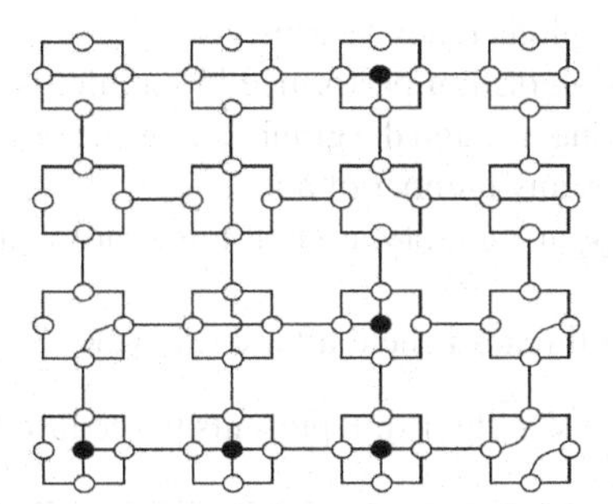

Fig. 1 R-Mesh-ACO

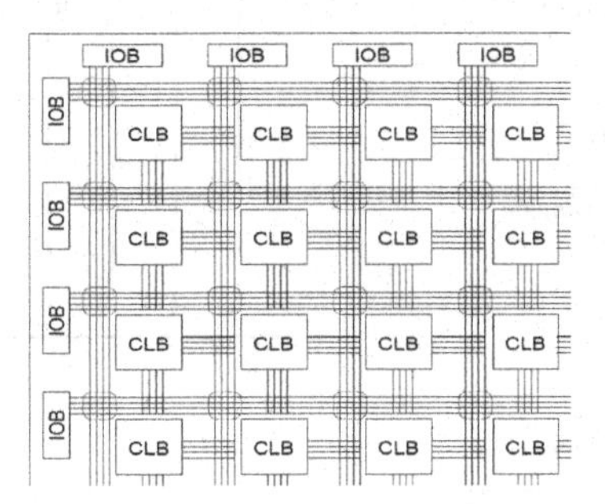

Fig. 2 FPGA-ACO

Most of the parallel implementations of the ACO known from the literature are only parallelization of standard ACO [14]. They differ only in details (e.g. number of ants served by a single processor) and the centralization or decentralization of the pheromone matrix (there are some exceptions — usually in multi-colony systems, where there are many colonies of ants, which have their own pheromone matrices and where each matrix may vary for different colonies). Very often, authors do not discuss in detail the results of calculations — they are probably not too good. The results obtained in the case of the deployment of individual ants on processors and the use of a central pheromone matrix show small performance gain due to the high cost of communication (about 12% time gain, 25%, 57% at 3, 5, 15 processors [28]). Some solutions even result in increased computing time [24]). The most vulnerable to parallelization are multi-colony systems.

6 The Parallel NAVN — PAVN

The usefulness of the NAVN algorithm has encouraged us to improve its performance by its parallelization [5]. Main features of the parallel version of NAVN (PAVN) are:

- the PAVN algorithm exploits many threads,
- the part of the NAVN algorithm performed in parallel is the same as in the sequential version (walking through the graph) and is performed by a simple thread,
- one thread may support any number of ants,
- threads use the common memory both for the problem description and the pheromone,
- the synchronization of threads is held after each cycle.

The pseudo-code of the PAVN algorithm presents procedure PAVNProc.

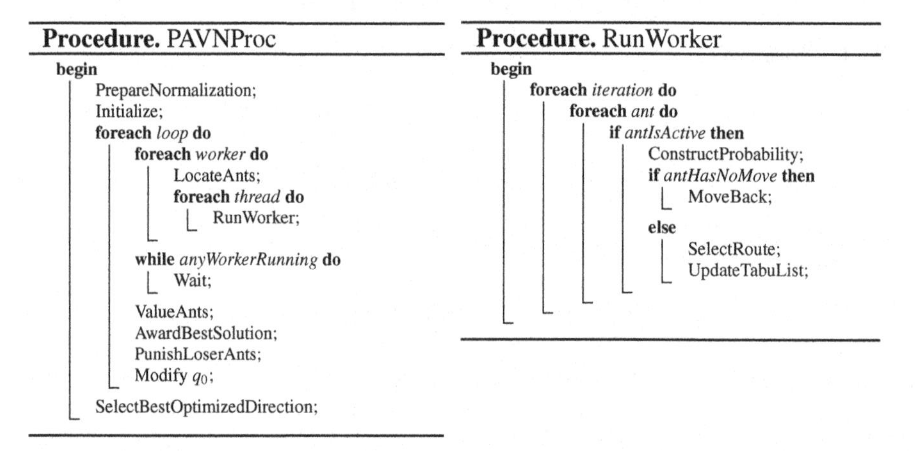

```
Procedure. PAVNProc
begin
    PrepareNormalization;
    Initialize;
    foreach loop do
        foreach worker do
            LocateAnts;
            foreach thread do
                RunWorker;
        while anyWorkerRunning do
            Wait;
        ValueAnts;
        AwardBestSolution;
        PunishLoserAnts;
        Modify q0;
    SelectBestOptimizedDirection;
```

```
Procedure. RunWorker
begin
    foreach iteration do
        foreach ant do
            if antIsActive then
                ConstructProbability;
                if antHasNoMove then
                    MoveBack;
                else
                    SelectRoute;
                    UpdateTabuList;
```

7 OpenCL Ant Vehicle Navigation Algorithm

Our further improvement of the PAVN algorithm was performed exploiting the computational potential of modern GPUs — devices equipped with the CUDA architecture (Compute Unified Device Architecture) from nVIDIA [19].

7.1 *The CUDA Architecture*

CUDA is an universal architecture, developed by nVIDIA, for parallel computing on graphics processing units (GPUs) and on dedicated accelerators (e.g. Tesla). This makes it possible to use GPUs for general computing performed so far only on CPUs (fig. 3). In contrast to the CPU, GPU's architecture requires the allocation of tasks to multiple threads which are executed relatively slowly, but globally offering a solution faster than traditional, general purpose, sequential processors. Such GPU devices are ideally suited for computations that can be run in parallel.

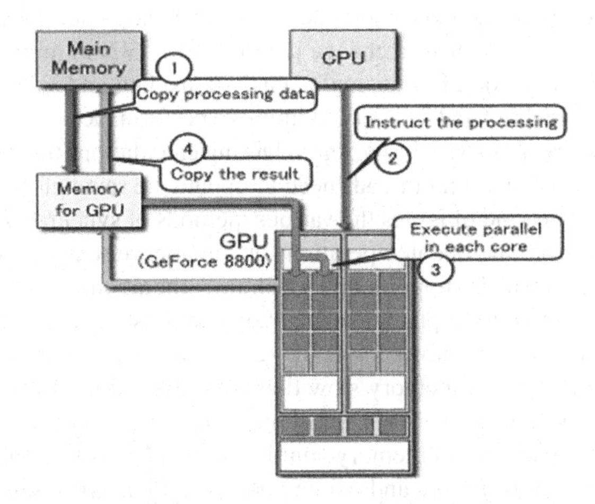

Fig. 3 Processing flow on CUDA

In order to be able to run our software on hardware from other manufacturers in the future, we decided to use an OpenCL programming interface. OpenCL (Open Computing Language) is a low-level API for heterogeneous computers, which may include the use of the CUDA architecture. Exploiting OpenCL, developers can create programs that run on the GPU device using programming language based on the C99 standard.

7.2 *Adaptation of the NAVN Algorithm for CUDA and OpenCL*

To be able to run the NAVN algorithm on GPU it was necessary to introduce various modifications to take into account the specificities of the target hardware architecture (CUDA) and the OpenCL programming interface as well. This resulted in the new version of the PAVN algorithm — OCLAVN. Below we present the basic characteristic of this approach.

Optimization of data transfer between host and device. Computations in CUDA architecture are performed on heterogeneous environment. The user application is divided into two parts: a host system code, and a GPU code, and both have to communicate each other. In OpenCL, data must be transferred from the host to the device, and vice verse. These transfers can decrease overall performance so they should be minimized. Data should be kept on the device as long as possible. The OpenCL version of PAVN meets these considerations. Almost all operations are performed by various kernels on GPU with data kept in device's memory.

Global memory usage minimization. The OpenCL specification describes three kinds of memory: global, local and private. They differ in term of scope, capacity and access time. Global memory has highest capacity and is accessible by every thread, but is significantly the slowest. Local memory is shared across threads in the same group, and private memory is private to a thread and is not visible to another threads. In many places we had to change the logic of PAVN to minimize the need of access to global memory. For example, ants' status data and solutions build by the ants are all located in local memory to improve performance.

Eliminating the need for synchronization. During calculations there are moments when two (or more) concurrent threads need to update the same global memory location. To solve these kind of issues the various methods of synchronization may be used. For example, atomic update operations or semaphore scheme can be used to avid the conflicts. OpenCL supports integer atomic operations on global memory. Therefore, it is possible to implement semaphores and use them to avoid simultaneous access to the same memory points by various threads. Unfortunately, atomic operations involving global memory slow down the execution of entire kernel, so we have decided to change the PAVN algorithm so that there is no synchronization. There are two major sources of memory conflicts in PAVN: on-line pheromone trail update in procedure *SelectRoute* and whole procedure *PunishNonActiveAnts*. They have been excluded from the final OCLAVN algorithm version, and the pheromone trail is now updated in procedures *AntMoveBack* and *AwardBestSolution*.

Code optimizations for the SIMD architecture. For best performance threads should be running in groups of at least 32 elements (warp size), with total number of threads of thousands. Branches in the program code do not impact performance significantly, providing the way that each of 32 threads takes the same execution path (the Single Instruction Multiple Data execution model). In various places we have changed the PAVN algorithm to achieve as much as possible the same execution path of all threads in the same group.

Other considerations. OCLAVN uses single-precision floats because they provide the best performance on CUDA GPUs. For the same reason native runtime math operations are used. Functions using *native_functionName* map directly to the hardware level. They are faster but provide somewhat lower accuracy, yet enough good for our application.

The OCLAVN algorithm may be described as shows procedure OCLAVNProc (procedure *RunAnts* is invoked for kernel calculations):

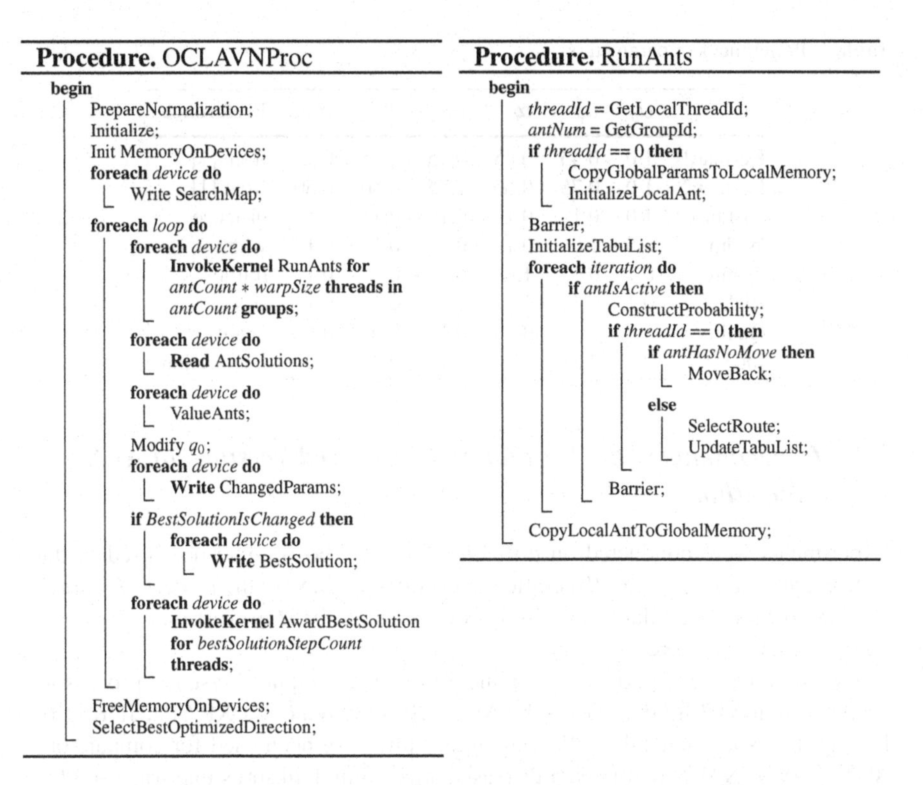

Procedure. OCLAVNProc

```
begin
  PrepareNormalization;
  Initialize;
  Init MemoryOnDevices;
  foreach device do
    Write SearchMap;
  foreach loop do
    foreach device do
      InvokeKernel RunAnts for
      antCount * warpSize threads in
      antCount groups;
    foreach device do
      Read AntSolutions;
    foreach device do
      ValueAnts;
    Modify q0;
    foreach device do
      Write ChangedParams;
    if BestSolutionIsChanged then
      foreach device do
        Write BestSolution;
    foreach device do
      InvokeKernel AwardBestSolution
      for bestSolutionStepCount
      threads;
  FreeMemoryOnDevices;
  SelectBestOptimizedDirection;
```

Procedure. RunAnts

```
begin
  threadId = GetLocalThreadId;
  antNum = GetGroupId;
  if threadId == 0 then
    CopyGlobalParamsToLocalMemory;
    InitializeLocalAnt;
  Barrier;
  InitializeTabuList;
  foreach iteration do
    if antIsActive then
      ConstructProbability;
      if threadId == 0 then
        if antHasNoMove then
          MoveBack;
        else
          SelectRoute;
          UpdateTabuList;
      Barrier;
  CopyLocalAntToGlobalMemory;
```

In the OCLAVN algorithm's pseudo-code special comment should be provided for keywords **Write**, **Read** and **InvokeKernel**:

Write realizes memory transfer from host to global memory on device,
Read — memory transfer from global memory on device to host memory, and
InvokeKernel runs given procedure on device by given number of threads (work items) divided into groups (work groups).

8 Computational Experiments

To demonstrate the proper operation of the proposed algorithms, several experiments have been carried out. The main objective was to obtain feasible solutions on the real world map in a satisfactory time. In the case of parallel versions of the algorithm the main goal was to proof that the runtime of the algorithms scale well with growing number of computing resources (number of CPU, CUDA cores in case of GPU).

Table 1 Preferences of parameters

	α_d	α_w	α_t	α_r	α_q	α_i	**Preference**
Kerman1	1.0	0.45	0.45	0.25	0.25	0.3	from [1]
Kerman2	1.0	0.75	0.60	0.75	0.50	0.50	from [1]
Distance	1.0	0.1	0.1	0.1	0.1	0.1	distance
Width	0.5	1.0	0.1	0.1	0.1	0.1	width
Traffic	0.1	0.1	1.0	0.1	0.1	0.1	traffic
Risk	0.1	0.1	0.1	1.0	0.1	0.1	risk
Quality	0.1	0.1	0.1	0.1	1.0	0.1	quality
Inter	0.1	0.1	0.1	0.1	0.1	1.0	intersection

8.1 Experiments with Original and Improved Versions of AVN Algorithm

Experiments were conducted on different data sets: (a) original data used in the experiment „*Kerman*" [25], (b) crafted data corresponding to the project „*London*" [26] and (c) large real data from the system *Open Street Map* (`http://www.o-penstreetmap.org/`).

Experiments were performed on the following computer system: processor: AMD Athlon 64 1.8 GHz 3000+, RAM: 1 GB, OS: Windows XP SP2, JDK: 1.6. During the experiments the following algorithms have been used for comparison: AVN, CAVN, NAVN and Dijkstra's classic algorithm. Dijkstra's one was used for comparison purposes only. It is not able to find a set of solutions, and therefore this algorithm is not interesting in scope of multi-criteria optimization.

The algorithms were run with different settings of preferences (tab.1) and with their parameters, including those for the ant algorithm, defined in tab.2 (m is the number of ants and $N_{\max}$ — loop count). The values of parameters of the ant algorithm were, as usual, set as a result of preliminary experiments.

8.1.1 Experiments on Data from the Project "Kerman" and "London"

Data for the experiments were obtained from the authors of the AVN algorithm [25, 26, 27]. Graph consists of 27 nodes and 67 edges, and represent partial map of the city of Kerman fig.4. The average number of edges (incoming and outgoing) per node is 2.48. The route was sought from point 8 to 22. Time of departure was 17:30 and average speed — 40 km/h.

Table 2 Parameters of the algorithms

Parameter	α	β	av	pv	bv	ρ	Q	φ	τ_0	m	$N_{\max}$
AVN	2	–	950	0.9	–	0.9	0.9	–	–	10	50
CAVN	2	–	1.05	0.9	–	0.9	0.9	0.99	–	10	50
NAVN	–	2.5	–	0.9	0.1	0.2	0.9	0.99	1/6000	10	50

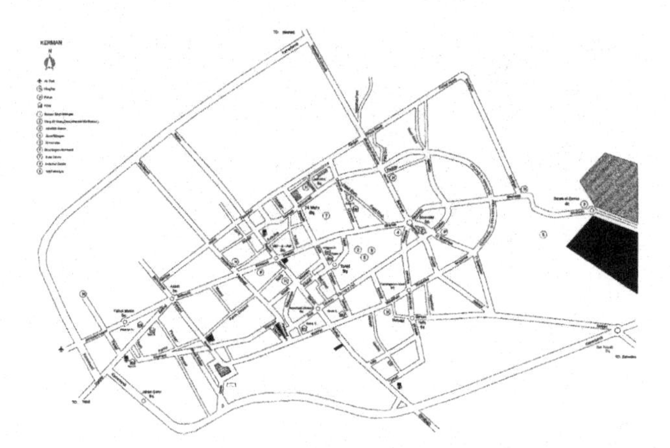

Fig. 4 Fragment of Kerman city map used in the experiments [1]

Algorithms were searching for a solution with the lowest possible cost. All algorithms, except AVN, found optimal solutions. Execution time of ant algorithms was greater than execution time of the deterministic algorithms.

Data for the experiment "London" were prepared so that the number of nodes was comparable with the data described in [26] (the original data were not available). Map consisted of 54 nodes and 184 edges. The average number of edges (incoming and outgoing) per node was 3.4. The route was sought from point 1 to 54. Time of departure was 17:30 and average speed — 40 km/h.

In this experiment the NAVN algorithm found the optimal solution, the same one as the deterministic one. Both AVN and CAVN algorithms did not find this solution, but CAVN produced better results. Dijkstra's algorithm was working at the non-measurable time. The overall results of experiments are presented in tab.3 (time is expressed in milliseconds).

8.1.2 Experiments on the Real Data

The data for this experiment were collected from the system Open Street Map (OSM) for the area of the city of Katowice (fig.5). Map size used in this experiment exceeds several times the size of the maps used in the projects "Kerman" and "London". This map consists of 31044 nodes and 68934 edges. The average

Table 3 Results of experiments with data of "Kerman" and "London"

	Kerman			London		
Algorithm	Time	Cost	Route	Time	Cost	Route
AVN	156	6290	8,2,5,4,6,7,3, 24,22	266	1185	1,3,6,7,8,9,11,28,30,31,29,32,34,33,40, 44,45,46,51,54
CAVN	79	5425	8,2,20,1,26,22	188	880	1,4,7,8,10,14,15,19,39,43,47,52,53,54
Dijkstra	≈0	5425	8,2,20,1,26,22	≈0	860	1,3,6,13,14,15,19,39,43,47,52,53,54

Fig. 5 Map used in experiments on real data

number of edges (incoming and outgoing) per node is 2.2. In this experiment only the preferences of the distance and width were used, other values were constant. This is due to the limited scope of information available in the OSM system. The start node was OSM id.: 383783583 (suburbs of Katowice) and end node id. was 384912139 (downtown of Katowice). The time of departure was 17:30 and travel speed: 40 km/h.

Table 4 Results of experiments with data of "Katowice"

	Distance			Width		
	Time	Cost	Distance	Time	Cost	Distance
NAVN	9531	38705	15466	11984	298984	19180
Dijkstra	2515	27144	13495	2531	212459	14317

During the preliminary experiments we found that the original algorithm AVN and its improved version (CAVN) were unable to find any solution, so they are omitted from presented results. Therefore the NAVN algorithm was compared only with the classical Dijkstra's algorithm. Results of experiments for the project "Katowice" presents tab.4.

The results show, that the NAVN algorithm can find good (although not optimal) solutions for large real map. This is important information considering the fact that the original algorithms AVN and CAVN do not give any solutions for such large data. Another important observation is that the execution time of the NAVN algorithm starts to be comparable to Dijkstra's algorithm. During some experiments this time was even shorter. Observing the trend of the execution time of both algorithms, we can risk the hypothesis that with further increasing the complexity of the maps, the NAVN algorithm may be faster than Dijkstra, of course, assuming a certain acceptable inaccuracy of solutions found. Especially interesting seems to be the result of the NAVN algorithm for preferences of width. The solution is regarded by many people (driving between the mentioned points of Katowice) as the best route, based mainly on the expressways (one of the authors uses the same one). The Dijkstra's algorithm, although it calculates the cost in exactly the same way, suggests a different route.

8.2 Experiments with the Parallel Version of AVN — PAVN

Tab. 6 presents selected results of experiments with NAVN and PAVN algorithms. All the versions were implemented in Java language and run on machine equipped with quad core processor (Intel Core2 Quad 2.4 GHz). This map consists of 64K nodes and 144K edges and is based on data collected from the system Open Street Map (OSM) for the area of the city of Katowice. The start node OSM id was 262831991 (city of Gliwice, Akademicka street) and end node id was 297573921 (city of Oświęcim, Zatorska street). The time of departure was 17:30 and travel speed: 40 km/h.

Table 5 NAVN and PAVN algorithm's parameters used during computations

Algorithm	α	β	Pv	Bv	ρ	Q	φ	τ_0
NAVN/PAVN	1.0	3.0	0.95	0.15	0.2	0.9	0.995	0.000025

Table 6 Selected results of experiments with NAVN i PAVN

Algorithm	Threads	Cycles	Ants	Time [ms]	Cost	Distance [m]
NAVN	1	50	16	20611	2290931	67055
NAVN	1	50	32	36723	2205954	68531
NAVN	1	100	32	49597	2277718	69556
PAVN	16	50	16	22033	2352790	81289
PAVN	8	50	16	13626	2426367	66374
PAVN	8	100	16	7766	2668998	83915
PAVN	8	50	32	13470	2737325	68416
PAVN	8	100	32	14673	2410930	81990

In experiment only the preferences of the distance were used, other values were constant. This is due to the limited scope of information available in the OSM system. Tab. 5 shows parameters' values, common to both algorithms used during experiments.

The results obtained by the PAVN algorithm in comparison with the sequential version of the algorithm (NAVN) were:

- The PAVN algorithm with the same number of cycles produced a little bit better results and the performance time was shorter (about 50%) for PAVN.
- Increasing the number of cycles led to further improvement of the results (with the performance time similar to the sequential version).
- Increasing the number of threads (the number of processor cores were constant) as well as the number of ants made both the computation time and the quality of results worse.

8.3 Experiments with CUDA Approach

Data for the experiments were also collected from the system Open Street Map (OSM) in the area of the city of Katowice. The map used in experiments consists of 31044 nodes and 68934 edges. The average number of edges (incoming and outgoing) per node is 2.2. The detailed data for the experiments were (other parameters of the OCLAVN are presented in Tab. 7): OSM start node id: 383783583 (suburbs of Katowice), OSM end node id: 384912139 (downtown of Katowice), departure time: 17:30, and speed: 40 km/h.

Table 7 OCLAVN algorithm's parameters used during computations

Algorithm	α	β	Pv	Bv	ρ	Q	φ	τ_0	m (no. of ants)	N_{max}
NAVN	1.0	2.0	0.9	0.1	0.2	0.9	0.995	1/44500	30	50
OCLAVN	0.1	1.0	0.9	0.1	0.1	0.9	0.995	1/44500	30	50

In experiments (as in previous ones) only the preferences of the distance were used, other values were constant. Tab. 7 shows parameters' values, common to both algorithms used during experiments. Changes in the logic of the original NAVN algorithm forced the usage of different parameters' values for sequential NAVN and parallel OCLAVN. These values have been discovered during preliminary runs of algorithms. Experiments were carried out on Microsoft Windows XP and, in case of GF GTX 295, on Windows 7. All the applications were compiled by Visual C++ 2008. OpenCL version of the algorithm was build with libraries from nVIDIA GPU Computing Toolkit 3.2. Sequential NAVN also uses single-precision floats, so that the experimental results were more reliable.

OpenCL version of the algorithm was run on various CUDA capable GPUs form nVIDIA, which are described in Tab. 8. The experiments allowed us to observe how hardware capabilities influence the results of the experiments. In case of GF GTX 295 only one device available on GPU was used.

Tab. 9 presents selected results of experiments on different platforms. The results shown are average values from 10 runs. Quality of each result (in term of lower cost value) is very similar. The OCLAVN run on GF GTX 295 has the best time performance, better than results obtained by sequential NAVN (in fact, the performance

Table 8 Parameters of selected nVIDIA GPUs

GPU	Multiprocessors	CUDA Cores	Compute Capability
GF 8400 GS	1	8	1.1
GF 9500 GT	4	32	1.1
Quadro NVS 320M	4	32	1.1
GF GTX 295	2x30	2x240	1.3

Table 9 Selected results of experiments with OCLAVN

CPU	GPU	Algorithm	Cost	Time (ms)
Intel Core2 Quad 2.4 GHz	GF 8400 GS	OCLNAVN	15702	39842
Intel Core2 Quad 2.4 GHz	GF 9500 GT	OCLNAVN	15152	18618
Intel Core2 Duo 2.5 GHz	Quadro NVS 320M	OCLNAVN	14939	15270
Intel Core I5 2.67 GHz	GF GTX 295	OCLNAVN	15538	**5232**
Intel Core I5 2.67 GHz	-	NAVN	15651	6424

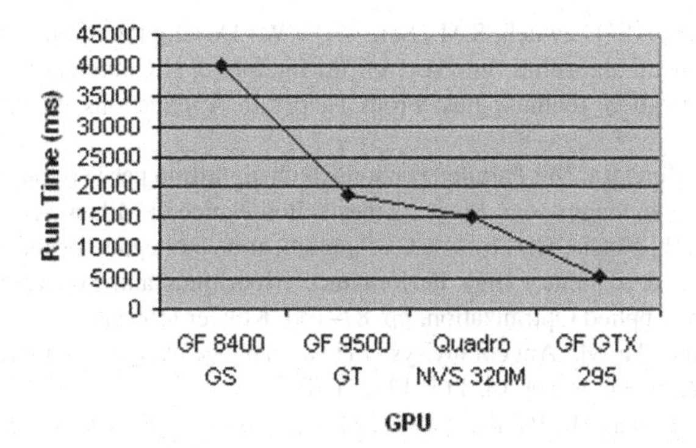

Fig. 6 OCLAVN algorithm runtime on different GPUs

on the CPU was only for the comparison of the quality of results, because features of the CPUs and GPUs for the sequential computations are incomparable). Fig. 6 presents the execution time of OCLAVN on different GPUs and it shows that the algorithm's performance scales with growing number of multiprocessors. It is not a linear speedup, because the base AVN is not a completely data parallel algorithm but it shows that the OCLAVN algorithm is well designed for the use of GPUs. The search space is located in global memory on GPU, so that the performance limitation of that kind of memory was the most important challenge for our algorithm's tuning attempts. Influence of the global memory performance limitations on overall algorithm performance can be decreased in the latest and future releases of GPUs. The promising example is new GPU architecture from nVIDIA called Fermi. One of the many improvements of the new architecture is significant speedup of global memory access.

9 Conclusions

The presented results of work on multi-agent ACO algorithms in car navigation shows that this is a promising line of research.

Ant colony optimization algorithms have a natural ability to solve problems of determining the optimal path in the graph, not only in case of single objective version, but also for multi-criteria problems.

Particularly noteworthy is the fact that the presented approach is able to construct a set of routes, not only "the best" one. This feature of the algorithm is particulary important for solving multi-criteria optimization problems, where the task is to find Pareto optimal set of solutions.

The presented parallel versions of the algorithm show high susceptibility of the sequential version of the algorithm to parallelization, which is important at the current trends in the development of the modern computing architectures.

References

1. Alhalabi, S.M., Al-Qatawneh, S.M., Samawi, V.W.: Developing a route navigation system using genetic algorithm. In: Proc. of 3rd International Conference on Information and Communication Technologies: From Theory to Applications, Damascus, pp. 1–6 (2008)
2. Bolondi, M., Bondaza, M.: Parallelizzazione di un algoritmo per la risoluzione del problema del comesso viaggiatore. Master's thesis, Politecnico di Milano (1993)
3. Bullnheimer, B., Kotsis, G., Strauss, C.: Parallelization strategies for the ant system. In: De Leone, R., et al. (eds.) High Performance Algorithms and Software in Nonlinear Optimization. Applied Optimization, pp. 87–100. Kluwer (1998)
4. Bura, W., Boryczka, M.: Ant colony system in ambulance navigation. Journal of Medical Informatics & Technologies 15, 115–124 (2010)
5. Bura, W., Boryczka, M.: Parallel version of the NAVN system (in polish). In: Systemy Wspomagania Decyzji, Uniwersytet Śląski w Katowicach (2010)
6. Colorni, A., Dorigo, M., Maniezzo, V.: Distributed Optimization by Ant Colonies. In: Proceedings of ECAL 1991 - European Conference on Artificial Life, pp. 134–142. Elsevier Publishing, Paris (1991)
7. Cordon, O., Herrera, F., Stutzle, T.: A review on the ant colony optimization mtaheuristic: Basis, models and new trends. Mathware & Soft Computing 9, 1–36 (2002)
8. Dorigo, M.: Parallel ant system: An experimental study. Politechnico di Milano (1993) (unpublished)
9. Dorigo, M., Di Caro, G.: The Ant Colony Optimization meta-heuristic. In: Corne, D., Dorigo, M., Glover, F. (eds.) New Ideas in Optimization, pp. 11–32. McGraw Hill, London (1999)
10. Dorigo, M., Di Caro, G., Gambardella, L.M.: Ant algorithms for discrete optimization. Artificial Life 5(2), 137–172 (1999)
11. Dorigo, M., Gambardella, L.M.: Ant Colony System: A cooperative learning approach to the Traveling Salesman Problem. IEEE Transactions on Evolutionary Computation 1, 53–66 (1997)
12. Garey, J.: Computers and Intractability: A Guide to the Theory of NPcompleteness. Freeman, San Francisco (1979)

13. Hansen, P.: Bicriterion path problems. In: Fandel, G., Gal, T. (eds.) Multiple Criteria Decision Making: Theory and Application. LNEMS, pp. 109–127. Springer, Heidelberg (1980)
14. Janson, S., Merkle, D., Middendorf, M.: Parallel ant algorithms. In: Alba, E. (ed.) Parallel Metaheuristics. Wiley Book Series on Parallel and Distributed Computing, pp. 171–201. Wiley (2005)
15. Kambayashi, Y., Tsujimura, Y., Yamachi, H., Yamamoto, H.: Integrating uncomfortable intersection-turns to subjectively optimal route selection using genetic algorithm. In: Proceedings of the 5th IEEE International Conference on Computational Cybernetics (ICCC 2007), Tunisia, pp. 203–208 (2007)
16. Krüger, F., Merkle, D., Middendorf, M.: Studies on a parallel ant system for the bsp model (1998) (unpublished)
17. Michels, R., Middendorf, M.: An ant system for the shortest common supersequence problem. In: Glover, F., Corne, D., Dorigo, M. (eds.) New Ideas in Optimization, pp. 51–61. McGraw-Hill (1999)
18. Middendorf, M., Reischle, F., Schmeck, H.: Multi colony ant algorithms. Journal of Heuristics 8(3), 305–320 (2002)
19. nVidia. What is CUDA, `http://developer.nvidia.com/what-cuda`
20. OpenStreetMap, `http://www.openstreetmap.org/`
21. Pang, G., Takahashi, K., Yokota, T., Takenaga, H.: Adaptive route selection for dynamic route guidance system based on fuzzy-neural approaches. IEEE Trans. on Vehicular Technology 48, 2028–2041 (1999)
22. Pangilinan, J.M.A., Janssens, G.K.: Evolutionary Algorithms for the Multiobjective Shortest Path Problem. In: World Academy of Science, Engineering and Technology, vol. 25 (2007)
23. Piriyakumar, D.A.L., Levi, P.: A new approach to exploiting parallelism in ant colony optimization. In: Proceedings of the International Symposium on Micromechatronics and Human Science (MHS), pp. 237–243 (2002)
24. Randall, M., Lewis, A.: A Parallel Implementation of Ant Colony Optimisation (2002)
25. Salehinejad, H., Farrahi-Moghaddam, F.: An ant based algorithm approach to vehicle navigation. In: Proceedings of the First Joint Congress on Fuzzy and Intelligent Systems (2007)
26. Salehinejad, H., Pouladi, F., Talebi, S.: A new route selection system: Multi-parameter ant algorithm based vehicle navigation approach. In: CIMCA 2008, IAWTIC 2008 and ISE 2008 (2008)
27. Salehinejad, H., Talebi, S.: A new ant algorithm based vehicle navigation system: A wireless networking approach. In: Proceedings of the International Symposium on Telecommunications (2008)
28. Souto, R.P., de Campos Velho, H.F., Stephany, S.: Reconstruction of chlorophyll concentration profile in offshore ocean water using a parallel ant colony code. In: Proc. 16th European Conference on Artificial Intelligence (ECAI), Hybrid Metaheuristics, HM (2004)
29. Teodorovic, D., Kikuchi, S.: Transportation route choice model using fuzzy interface technique. In: Proc. 1st International Symposium on Uncertainty Modeling and Analysis: Fuzzy Reasoning, Probabilistic Models, and Risk Management, pp. 140–145. Univ. Maryland, College Park (1990)
30. You, Y.-S.: Parallel ant system for traveling salesman problem on GPUs (2010) (unpublished)

Solving Instances of the Capacitated Vehicle Routing Problem Using Multi-agent Non-distributed and Distributed Environment

Dariusz Barbucha

Abstract. Applying metaheuristics to solve large scale instances of computationally difficult optimization problems often requires using a considerable computational effort in order to reach the satisfactory results in reasonable amount of time. Parallel/distributed computation may improve performance of such approaches. It is expected that parallel metaheuristics will outperform their sequential counterparts in terms of quality of the generated solutions as well as reducing the computation time. Last years, an agent paradigm has emerged as an interesting direction for effective solving different problems. The chapter focuses on multi-agent system JABAT, dedicated for solving computationally hard optimization problems using parallel and distributed environment. Two models of computations used by JABAT, where all software agents are running on the one container and where selected software agents are distributed (moved or cloned) over available additional containers (nodes), are presented in the chapter. The influence on the above models on quality of the results and the computation time has been investigated by computational experiment, which has been carried out on selected instances of capacitated vehicle routing problem.

1 Introduction

Solving large scale instances of computationally difficult optimization problems often require using a considerable computational effort in order to reach the satisfactory results in reasonable amount of time. The nature of metaheuristics used for solving such problems as well as some specific features of the problems give the possibility of using parallel computation. Parallel/distributed computing is a form of computation in which different processes work simultaneously on several processors/machines solving a given problem instance. Parallelism thus follows from

Dariusz Barbucha
Department of Information Systems, Gdynia Maritime University,
Morska 83, 81-225 Gdynia, Poland
e-mail: barbucha@am.gdynia.pl

I. Czarnowski et al. (Eds.): Agent-Based Optimization, SCI 456, pp. 55–75.
DOI: 10.1007/978-3-642-34097-0_3

a decomposition of the total computational load and the distribution of the resulting tasks to available processors. As Crainic and Nourredine [9] report, in case of metaheuristic methods such decomposition may concern the algorithm (functional parallelism), or the problem-instance data (data parallelism or domain decomposition). In functional parallelism, different tasks are allocated to different processors and run in parallel, possibly working on the same data and/or exchanging information. Data parallelism (or domain decomposition) refers to the case where the problem domain (or the associated search space), is decomposed and particular solution methods are used to address the problem on each of the resulting components of the search space.

The main reason for using parallel and distributed computing in the design and implementation of metaheuristics is to speed up the search. Reducing the search time is specially important in case of methods which are proposed to be used for solving complex optimization problems where the search time is critical (for example dynamic problems), or for solving large-scale instances of these problems. Very important feature of parallel implementations of metaheuristics is also ability to improve the quality of solutions obtained by these methods in comparison to the sequential implementations. Parallel metaheuristics may be also more robust than their sequential counterparts in terms of solving different optimization problems and different instances of a given problem in an more effective manner [29].

In the recent years, technological advances enabled development of various parallel and distributed versions of metaheuristics based on the multi-agent paradigm and using the agent technology. The term agent (or software agent), has found its way into a number of technologies and has been widely used, for example, in artificial intelligence and distributed computing areas. Although, as Jennings et al. [16] point out, there is no real agreement what an agent is, many authors involved in agent research have offered a variety of definitions, in which they explain their understanding of term 'agent' and emphasize its main features. For example, Jennings et al. [16] define agent as a computer system, situated in some environment, that is capable of flexible autonomous action in order to meet its design objectives [16]. According to definition of Russel and Norvig [26], an agent is anything that can be viewed as perceiving its environment through sensors and acting upon that environment through effectors. Another definition says that autonomous agents are computational systems that inhabit some complex dynamic environment, sense and act autonomously in this environment. By doing so they realize a set of goals or tasks for which they are designed [20]. And finally, definition provided by IBM says that intelligent agents are software entities that carry out some set of operations on behalf of a user or another program, with some degree of independence or autonomy, and in so doing, employ some knowledge or representation of the user's goals or desires [14].

Several interesting features possessing by agents (some of them occur in the above definitions) decide about the still growing interest in the exploration of the agent technology and application it to various fields. First, an agent is *autonomous*, because it operates without the direct intervention of humans or others and has control over its actions and internal state. An agent is *social*, because it cooperates with humans or other agents in order to achieve its tasks. An agent is *reactive*, because it perceives its environment and responds in a timely fashion to changes that occur in the environment. An agent is *proactive*, because it does not simply act in response to its environment but is able to exhibit goal-directed behavior by taking initiative [7]. Moreover, if necessary a software agent can be *mobile*, which means that it has ability to migrate or clone in the computer network [24] [31]. Mobile agents can be effectively used in various areas, including improvements in latency and bandwidth of client-server applications as well as reducing vulnerability to network disconnection. By using mobile agents the system allows for decentralization of computation processes and balancing of the load. This results in a more effective use of the available resources and reduction of the computation time.

Despite the above important agent's features, using a single agent in complex real-life applications is often not effective, so multi-agent systems (MAS) composed of multiple autonomous components (agents) are used [16]. Such multi-agent systems can model complex systems and introduce the possibility of agents having common or conflicting goals. Moreover, these agents may interact with each other both indirectly (by acting on the environment) or directly (via communication and negotiation). Also, agents may decide to cooperate for mutual benefit or may compete to serve their own interests [7].

Nowadays agent technology is used to solve real-world problems in a range of industrial and commercial applications. In a number of such approaches, the agent technology is integrated with different search paradigms, like for example metaheuristics. One example of such an approach is a concept of the asynchronous team (A-Team) [30], which integrates paradigms of the population-based methods, cooperative problem solving, parallel/distributed computing and multi-agent systems. Generally, A-Team can be characterized as a collection of software agents which cooperate to solve a problem by dynamically evolving the population of solutions stored in the common memory.

One of the approach based on the concept of A-Team, is a middleware called JABAT (JADE-Based A-Team) [3], dedicated for solving computationally hard optimization problems, which has been developed as a team work, with contribution of the author of this chapter. This chapter focuses on JABAT ability to distribute computation load while system is solving instances of given optimization problem. The main goal is to investigate the influence of the distribution of the computation process, where selected software agents are distributed (moved or cloned) over available platforms (nodes), on quality of the results and the computation time. The reported investigation has been carried out using several selected instances of Capacitated Vehicle Routing Problem.

The rest of the chapter is organized as follows. Section 2 addresses the parallel metaheuristic methods, presents their classification and overview of their implementations for vehicle routing problem. Section 3 presents an overview of the JABAT system, its features, elements of architecture, and implementation details. A goal, assumptions, and results of the computational experiment are included in Section 4. And finally, Section 5 presents concluding remarks, and highlights directions for future research.

2 Parallel Metaheuristics

The most exhaustive classification of parallel strategies for metaheuristics, in the opinion of the author, has been proposed by Crainic and Nourredine [8], which next has been adopted and extended by Crainic and Toulouse [9]. Three dimensions of the classification have been distinguished.

The first dimension (*Search Control Cardinality*) indicate how the global problem solving process is controlled. It may be controlled by a single process or by several processes that may collaborate or not. These two alternatives are identified as 1-control (*1C*) and p-control (*pC*), respectively.

The second dimension (*Search Control and Communications*) addresses the issue of information exchanges. In parallel computing, one generally refers to synchronous and asynchronous communications. In the first form of communication, all processes stop and engage in some form of communication and information exchange at moments determined by a features of algorithms or by master process (number of iterations, time intervals, specified algorithmic stages, etc.). On the other hand, in asynchronous communication, each process is in charge of its own search, it often decides about establishing communications with other processes, and the global search terminates once each individual search stops. To reflect more adequately the quantity and quality of the information exchanged and shared, as well as the additional knowledge derived from these exchanges (if any) four classes are defined: Rigid (*RS*) and Knowledge Synchronization (*KS*) and, symmetrically, Collegial (*C*), and Knowledge Collegial (*KC*).

The third dimension (*Search Differentiation*) reflects the fact of starting the search threads from the same or different solutions and making use of the same or different search strategies. The four cases are considered here: Same initial Point/Population, Same search Strategy (*SPSS*), Same initial Point/Population, Different search Strategies (*SPDS*), Multiple initial Points/Populations, Same search Strategies (*MPSS*), and Multiple initial Points/Populations, Different search Strategies (*MPDS*).

Basing on the above classification and the sources of parallelism in meta-heuristics, Crainic and Toulouse [9] identified four groups of the parallel meta-heuristic strategies: 1-control strategies, strategies based on explicit domain decomposition, independent multi-search strategies, and cooperative multi-search strategies.

The first group of strategies (1-control strategies), exploits the intrinsic parallelism offered by the basic, inner-loop, computations of metaheuristics, and are usually implemented according to the classical master-slave parallel programming model. Whereas a master program executes the 1-control metaheuristic, computation-intensive tasks are dispatched to slave programs in order to being executed in parallel by them. Taking into account the above taxonomy, they belong to the *1C/RS* class. An example of using this strategy is a tabu search implementation of Garcia et al. [13] for the Vehicle Routing Problem with Time Windows. In their parallel synchronous algorithm many different neighborhoods of the current solution are considered, and next manny modifications to the current solution are applied. Another example of implementation of this strategy is a parallel ant colony approach of Doerner et al. [10] for Capacitated Vehicle Routing Problem. Tasks, including one or several ants, is built by master process and next are solved by slave processes.

Domain (or search-space) decomposition is based on the idea of dividing the whole search space into smaller (usually disjoint) subspaces, solving the resulting subproblems (operating on these subspaces) by applying the sequential metaheuristic on each of them, collecting the respective partial solutions, and building the complete solution from these partial solutions. This strategy is usually represented by *1C/KS* schemes, with a *MPSS* or *MPDS* search differentiation strategy. In case of Vehicle Routing Problems, a set of customer may be initially clustered (a depot is included in each cluster), and next, several subproblems, may be solved independently by separated search threads running in parallel. Next, a final solution is a composition of these partial solutions which come from subproblems. An approach for Vehicle Routing Problem which belongs to this group is, for example [27], where a set of subproblems were solved by tabu-search metaheuristic.

Independent multi-search consists in performing several searches simultaneously on the entire search space, starting from the same or from different initial solutions, and selecting at the end the best among the best solutions obtained by all searches. Independent multi-search methods belong to the *pC/RS* class. Doerner et al. [11] studied different parallel implementations of the Savings based Ant System algorithm developed for solving the Vehicle Routing Problem. They analyze the effects of different form of parallelization (low-level parallelization, multiple search strategies and domain decomposition approaches). In case of multiple search strategies they investigated different information exchange schemes.

Cooperative multi-search methods, similar to the independent multi-search ones, initiate several simultaneous independent search threads, each defining a trajectory in the search space from a possibly different initial point or population by using a possibly different methods or search strategy. Hovewer, they implement also the information sharing cooperation mechanism specifying how these independent methods interact within the global search behavior of the cooperative parallel metaheuristic emerging from the local interactions among them. They belong to the *pC/KS* or *pC/C* or *pC/KC* groups of strategies. As an example of using coperative approaches for Vehicle Routing Problem one can list those proposed by Le Bouthillier et al. [18] and Meignan et al. [21]. Le Bouthillier et al. [18] proposed the framework for a guided parallel cooperative search for the Vehicle Routing Problem with Time Windows based on the central memory multi-thread cooperative search concept. Meignan et al. [21] proposed a coalition-based metaheuristic (CBM), merging some of the evolutionary algorithm concepts into a context of distributed control and agent-based learning.

Last years, a few agent-based approaches have been also proposed which implement metaheuristics procedures possibly to be executed in parallel. One of them is, for example, a coalition-based metaheuristic suggested by Meignan et al. [21] and mentioned above. Another, worth mentioning in this context, is a multi-agent architecture for metaheuristics MAGMA proposed by Milano and Roli [22]. A JABAT-based approach, considered in this chapter, also belongs to a group of approaches combining metaheuristic, cooperative problem solving and multi-agent paradigms. According to the taxonomy of Crainic and Toulouse [9], it may belong to *pC/C or KC/SPDS or DPDS* classes.

3 JADE-Based A-Team

A middleware called JABAT (JADE-Based A-Team), supporting design and implementation of multi-agent parallel architectures for solving difficult optimization problems is based on the concept of the asynchronous team (A-Team) [30]. A-Team is a collection of software agents (each of them encapsulates a particular problem-solving method) which collectively work and cooperate to solve a problem by dynamically evolving the population of solutions stored in the common memory. Within an A-Team, agents are autonomous and asynchronous, and each agent encapsulates a particular problem-solving method, which usually is inspired by natural phenomena like, for example, evolutionary processes or particle swarm optimization, as well as local search techniques. The ground principle of asynchronous teams rests on combining algorithms, which alone could be inept for the task, into effective problem-solving organizations, possibly creating a synergetic effect, in which the combined effect of cooperation between agents is greater than the sum of their separate effects.

JABAT is built using JAVA programming language and JADE agent platform (Java Agent Development Framework) - a software framework proposed by TILAB [15] for the development and run-time execution of peer-to-peer applications.

JADE is based on the agents paradigm in compliance with the FIPA (Foundation for Intelligent Physical Agents) [12] specifications and provides a comprehensive set of system services and agents necessary to implement distributed peer-to peer applications in the fixed and mobile environment. It includes both the libraries required to develop application agents and the run-time environment that provides the basic services and that must be running on the device before agents can be activated.

JADE also manages the whole agent life cycle, provides the transport mechanism and interface to send/receive messages to/from other agents and supports debugging, management and monitoring phases with using dedicated graphical tools. Moreover, it also supports agents migration, complex interaction protocols, messages content creation and management including XML and RDF [7].

Till now, several different versions of the JABAT system have been implemented: A-Team implementation [3], Web-based A-Team implementation [4], and island-based A-Team implementation [5]. Here, taking into account the goal of the chapter, only a short overview of the main part of these implementation is given, since the reader can find further details of them in the above papers.

3.1 Overview of JABAT

Main functionality of the proposed environment is organizing and conducting the process of search for the best solution. The search process includes initialization and improvement steps. At first, the initial population of solutions is generated. Next, individuals forming the initial population are improved by independently acting autonomous optimization agents. And finally, after reaching the stopping criterion, the best solution from the population is taken as the result.

Main components of the system include:

- three kinds of agents directly engaged in search (*Task Manager*, *Solution Manager*, and a set of *Optimizing Agents*),
- three kinds of agents which role is to monitor the process the search (*Solution Monitor*, *Error Monitor*, and *Platform Manager*),
- population of individuals,
- communication between them.

3.1.1 Agents Directly Engaged in the Process of Search

Task Manager agent runs first. It is responsible for reading all needed global system parameters stored in the configuration file. When *TaskManager* finds a request to solve an instance of the problem its role is also to create *Solution Manager* and *Optimizing Agents*, designated to this particular task instance. One of the important role of the *Task Manager* is also to administer several searches conducted in parallel, which are allowed in the proposed system.

Solution Manager and a set of *Optimizing Agents* are mainly engaged in organizing and conducting the process of search for the best solution. Each *SolutionManager* is responsible for generation of the initial pool of solutions, sending periodically solutions to the *Optimizing Agents* taken from the pool of solutions, merging improved solutions with the population and deciding when the whole process of searching for the best solution should be terminated.

Each *Optimizing Agent* is a single autonomous and improvement algorithm. During its functioning it communicates with *Solution Manager* by sending out the messages. At the beginning of the cycle, it sends a message about its readiness to work. In response the *Solution Manager* sends the details of the task and a solution to improve. The *Optimizing Agent* improves it and sends back the improved solution to the *Solution Manager*. The process iterates, until some stopping criterion is met.

3.1.2 Agents Monitoring the Process of Search

The main role of the *Solution Monitor* is to record computation results during the whole process of search. Such results include information about the solutions improved by *Optimizing Agents*, computation time, average fitness of the population, etc.

The second monitoring agent - *Error Monitor* monitors and reports unexpected behavior of the system during the whole process of search.

And the third agent belonging to this group - *Platform Manager* is engaged in organizing and conducting several parallel searches, which can be performed on a single computer or on multiple computers. In the former case, if JABAT is activated on a single computer with main container, then all agents are placed and running on this one computer. The latter case is made possible due to the functionality of JADE [7], which allows agents to migrate or clone to containers on other computers that have joined the main platform (the role of *Platform Manager* agent is organizing the process of agents migrations between different platforms).

Platform Manager, in the course of its life, monitors the number of running *Optimizing Agents*, periodically checks the number of available containers, evaluate the utilization of the available containers by *Optimizing Agents* and, if desired, in order to load balance on the available containers it moves agents to other locations.

3.2 Models of Computations

Two models of computations performed on a single container (node) or with using additional containers, available in JABAT, are presented in Fig. 1.

If JABAT is activated on a single computer (or node) with main container, then all agents are placed and running on this one computer. If JABAT has been activated on multiple computers (with additional nodes), with main container placed on one computer and the remote joined containers placed on other computers (nodes), then the main container hosts the managing agents including the *Task Manager*, *Solution Manager*, and *Platform Manager*, monitoring agents (*Solution Monitor*, *Error Monitor*) while *Optimizing Agents* are moved from the main container to available additional containers (nodes) to distribute the workload evenly.

4 Computational Experiment

The goal of the experiment was to compare two models of computations (with and without distribution of software agents) of the multi-agent system JABAT. Two measures have been chosen in order to evaluate the performance of the system: mean relative error (in %) from the optimal or best known solution and computation time (in sec.) used by the system in order to obtained the (sub-)optimal solution.

The reported experiment has been carried out on instances of the Capacitated Vehicle Routing Problem (CVRP). CVRP is the network optimization problem, in which a set of customers is to be served by the fleet of vehicles in order to minimize the service cost and satisfy several customers and vehicles constraints: each route starts and ends at the depot, each customer is serviced exactly once by a single vehicle, and the total load on any vehicle associated with a given route does not exceed vehicle capacity [17].

Elements, which should be defined in order to use the JABAT for solving CVRP (representation of an individual, size of the population and method of its creation, methods of management of population of solutions, optimizing agents representing

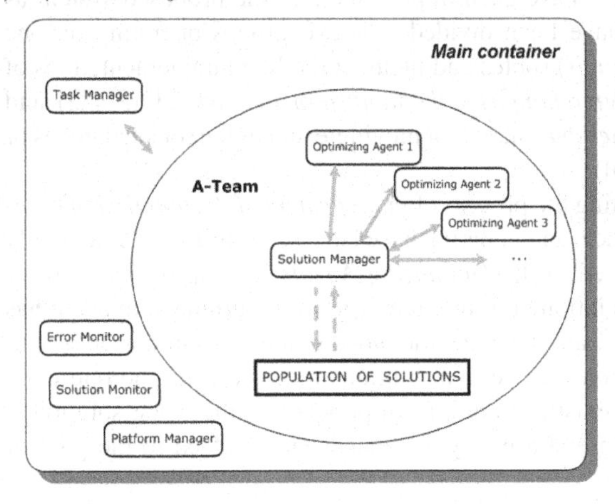

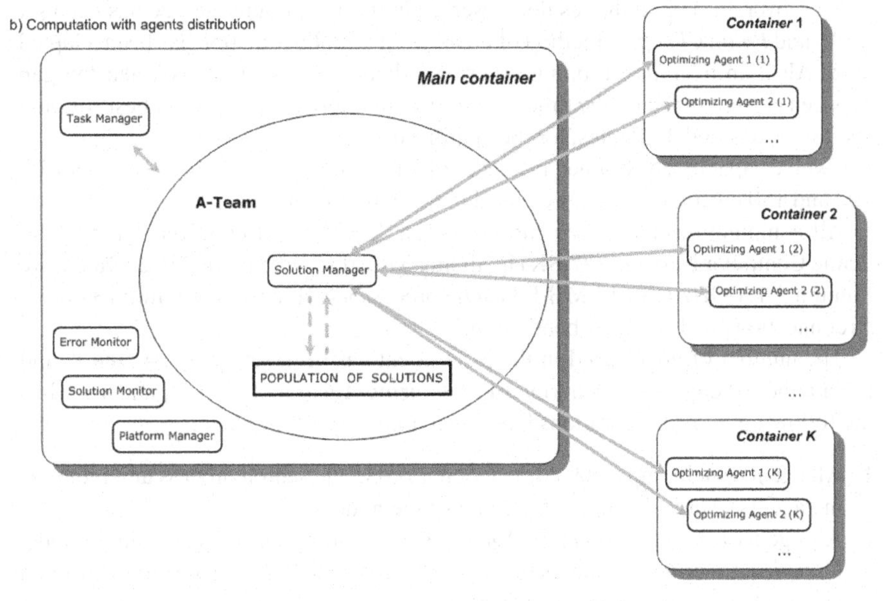

Fig. 1 Two models of computations on a single container or on multiple additional containers, respectively, available in JABAT

solution methods, and stopping criterion) have been adapted from [6], where the reader can find more details about them.

Path representation (a permutation of N numbers representing customers, where the order of numbers reflects the order in which customers are visited) known from Traveling Salesman Problem (TSP) has been adapted as *representation of individual*. Additionally, procedure of splitting the individual on segments (routes) has been implemented.

Four local search heuristics have been implemented in the proposed system as *optimization agents*. They have been divided into two groups operating on one (*intra-route*) or two (*inter-routes*) routes and include: modified implementations of *3-opt* procedure [19] and λ*-interchange local optimization* method [23] ($\lambda = 2$), and two dedicated *local search methods*, based on moving/exchanging selected nodes or edges between routes (see [6]).

Three operations constituting the process of *management of the population of individuals*: selection, acceptance, and updating have been set as follows. A randomly chosen individual is forwarded to the *Optimizing Agents* for improvement. After improvement phase, if the solution currently received from optimization agent has been improved, it is accepted and replaces the worst solution from current population. Additionally, if last consecutive five solutions received from the optimization agents did not improve existing solutions in population, the worst solution is removed from the population and a newly generated one is added to the pool of individuals.

Moreover, basing on the results presented in [6], the population size has been set to 30, and *Polar/Cheapest* method of creating an initial population has been adapted here. Also, an adaptive stopping criterion, where the system stops when the gap between current time and the time, where system improved the current best solution last time exceeded 3 minutes, has been utilized.

Twelve capacity constrained instances of VRP proposed by Taillard [28] with 75, 100 and 150 customers have been used in the experiment.

All computations have been carried out on the cluster Holk of the Tricity Academic Computer Network (TASK) including 34 nodes and built of 256 2xDual Core Itanium 2 1.4 GHz with 12 MB L3 cache processors with Mellanox InfiniBand interconnections with 10 Gb/s bandwidth.

The number of nodes used in the experiment has been set to 1, 2, 4, and 8, and the number of copies of each kind of *Optimizing Agent* to 1, 2, 4, 8, and 16. Thus the following two configurations have been used in the experiment:

1. All agents (including agents directly engaged in the search process and monitoring agents) were running on the same single node,
2. All agents (except *Optimizing Agents*) were running on the same single node, *Optimizing Agents* (nc copies of each, where $nc = 1,2,4,8,16$) were running on available additional nn nodes, where $nn = 2, 4, 8$.

Table 1 summarizes the above configurations and additionally presents total number of *Optimizing Agents* running in the system and the number of *Optimizing Agents* running on a single node.

Computational results are presented in Tables 2-6, separately for each team consisting of 1, 2, 4, 8, and 16 copies of each *Optimizing Agent*, respectively. The columns of each table include: instance name, mean relative error (separately for each number of nodes used in the experiment) and computation time (also separately for each number of nodes used in the experiment). Moreover, average values

Table 1 Tested cases including the number of *Optimizing Agents* engaged in search process and the number of nodes used by these agents

Number of copies of each Optimizing Agent	Number of nodes used	Total number of running agents	Number of agents running on single node
1	1	4	4
	2	4	2
	4	4	1
	8	4	1
2	1	8	8
	2	8	4
	4	8	2
	8	8	1
4	1	16	16
	2	16	8
	4	16	4
	8	16	2
8	1	32	32
	2	32	16
	4	32	8
	8	32	4
16	1	64	64
	2	64	32
	4	64	16
	8	64	8

Table 2 Mean relative error from the best known solution and computation time calculated for all instances and all considered number of nodes used by *Optimizing Agents* (1 copy of each agent)

Instance	Mean relative error [%]				Computation time [s]			
	1	2	4	8	1	2	4	8
tai75a	1.01	0.71	0.46	0.65	326.22	203.61	204.82	306.37
tai75b	0.79	0.52	1.00	0.85	245.13	214.60	167.85	256.58
tai75c	0.89	1.23	0.98	0.94	288.74	181.10	145.65	143.09
tai75d	0.34	0.65	0.45	0.47	273.96	191.74	215.93	238.81
Average(tai75)	0.76	0.78	0.72	0.73	283.51	197.76	183.56	236.21
tai100a	2.70	2.78	2.69	2.90	483.61	340.68	394.16	438.06
tai100b	0.72	0.68	0.77	0.67	452.33	291.52	293.82	353.16
tai100c	0.97	1.17	1.30	1.34	413.80	273.50	338.22	416.05
tai100d	2.34	2.07	2.36	2.09	391.11	350.43	286.93	474.65
Average(tai100)	1.68	1.67	1.78	1.75	435.21	314.03	328.29	420.48
tai150a	3.36	3.91	3.06	3.48	1066.82	682.84	690.40	955.13
tai150b	5.41	4.92	5.00	5.14	1190.98	969.33	960.64	912.98
tai150c	3.17	3.09	2.92	3.68	846.49	868.14	706.98	811.42
tai150d	2.76	2.29	2.65	2.55	977.85	825.06	873.99	970.27
Average(tai150)	3.68	3.55	3.41	3.71	1020.53	836.34	808.00	912.45
AVERAGE	2.04	2.00	1.97	2.06	579.75	449.38	439.95	523.05

Table 3 Mean relative error from the best known solution and computation time calculated for all instances and all considered number of nodes used by *Optimizing Agents* (2 copies of each agent)

Instance	Mean relative error [%]				Computation time [s]			
	1	2	4	8	1	2	4	8
tai75a	1.16	0.81	0.89	0.66	346.58	203.53	154.71	150.40
tai75b	0.74	0.94	1.01	1.01	389.48	206.35	108.08	117.67
tai75c	1.21	0.75	0.70	0.73	281.47	164.74	161.67	145.15
tai75d	0.83	0.77	0.40	0.22	289.99	194.85	163.49	125.97
Average(tai75)	0.98	0.82	0.75	0.65	326.88	192.37	146.99	134.80
tai100a	2.58	2.70	2.81	2.85	459.87	332.46	175.56	285.33
tai100b	1.17	0.91	1.16	0.92	397.72	259.85	217.48	168.56
tai100c	1.31	1.02	1.09	1.20	593.93	365.24	189.98	207.20
tai100d	2.22	2.23	2.17	2.88	425.85	243.68	185.20	205.41
Average(tai100)	1.82	1.72	1.81	1.96	469.34	300.31	192.05	216.63
tai150a	2.70	3.53	3.36	2.87	1273.08	637.64	718.49	592.66
tai150b	4.57	4.83	4.91	5.54	1289.73	945.36	618.35	638.50
tai150c	2.69	3.47	3.35	2.99	1111.04	840.65	563.10	676.52
tai150d	2.22	2.41	2.23	2.54	1008.18	727.49	635.06	556.46
Average(tai150)	3.04	3.56	3.46	3.48	1170.51	787.79	633.75	616.04
AVERAGE	1.95	2.03	2.01	2.03	655.58	426.82	324.26	322.49

Table 4 Mean relative error from the best known solution and computation time calculated for all instances and all considered number of nodes used by *Optimizing Agents* (4 copies of each agent)

Instance	Mean relative error [%]				Computation time [s]			
	1	2	4	8	1	2	4	8
tai75a	0.62	1.05	1.01	0.72	396.90	256.81	91.97	69.00
tai75b	0.91	1.08	0.78	0.57	353.36	265.50	164.15	98.48
tai75c	0.77	1.24	1.02	1.10	244.86	231.58	114.38	73.17
tai75d	0.36	0.49	0.55	0.48	354.70	259.74	125.45	111.55
Average(tai75)	0.67	0.96	0.84	0.72	337.46	253.41	123.99	88.05
tai100a	2.89	2.37	2.75	2.39	702.46	398.37	180.59	248.27
tai100b	0.61	0.71	1.21	0.72	504.38	377.80	160.36	126.64
tai100c	1.01	1.25	1.12	1.20	809.19	383.07	174.48	206.31
tai100d	2.19	2.27	1.59	1.88	568.71	382.53	222.00	152.17
Average(tai100)	1.67	1.65	1.67	1.55	646.18	385.44	184.36	183.35
tai150a	2.81	2.48	3.02	3.20	1513.23	941.66	586.11	346.34
tai150b	5.48	4.82	5.11	4.41	1446.19	1040.92	787.54	539.09
tai150c	2.94	2.74	2.45	3.13	1583.21	890.72	433.58	491.27
tai150d	2.65	2.55	2.01	2.36	1248.41	923.72	612.73	455.80
Average(tai150)	3.47	3.15	3.15	3.28	1447.76	949.26	604.99	458.12
AVERAGE	1.94	1.92	1.89	1.85	810.47	529.37	304.45	243.17

Table 5 Mean relative error from the best known solution and computation time calculated for all instances and all considered number of nodes used by *Optimizing Agents* (8 copies of each agent)

Instance	Mean relative error [%]				Computation time [s]			
	1	2	4	8	1	2	4	8
tai75a	0.60	0.54	0.78	0.87	402.31	334.92	162.81	78.33
tai75b	0.42	0.71	0.67	0.78	457.44	303.15	146.46	147.98
tai75c	0.92	0.50	0.93	1.35	447.01	285.29	111.23	51.57
tai75d	0.69	0.34	0.52	0.21	410.32	221.65	155.25	75.57
Average(tai75)	0.66	0.52	0.72	0.80	429.27	286.25	143.93	88.36
tai100a	2.46	2.25	2.10	2.15	698.35	502.40	256.74	165.88
tai100b	0.78	0.86	0.95	0.91	628.66	459.20	244.18	124.91
tai100c	0.94	0.95	1.25	1.03	651.16	389.67	222.09	99.64
tai100d	2.06	1.68	1.85	2.06	540.45	463.96	175.38	130.75
Average(tai100)	1.56	1.44	1.54	1.54	629.66	453.81	224.60	130.29
tai150a	2.47	2.65	2.60	2.95	1481.72	1057.69	608.44	400.62
tai150b	4.28	4.28	4.08	4.75	1651.00	1326.58	732.96	424.42
tai150c	2.95	2.93	2.92	2.60	1942.52	1270.65	625.73	472.78
tai150d	2.37	2.21	2.12	2.16	1535.63	1059.16	796.28	440.96
Average(tai150)	3.02	3.02	2.93	3.12	1652.72	1178.52	690.85	434.69
AVERAGE	1.75	1.66	1.73	1.82	903.88	639.53	353.13	217.78

Table 6 Mean relative error from the best known solution and computation time calculated for all instances and all considered number of nodes used by *Optimizing Agents* (16 copies of each agent)

Instance	Mean relative error [%]				Computation time [s]			
	1	2	4	8	1	2	4	8
tai75a	0.51	0.36	0.57	0.65	574.90	367.98	226.30	135.26
tai75b	0.67	0.36	0.50	0.38	560.28	404.51	188.73	180.85
tai75c	0.51	0.76	0.49	0.62	521.16	289.24	192.09	211.11
tai75d	0.22	0.14	0.41	0.07	520.23	401.76	222.18	149.62
Average(tai75)	0.48	0.40	0.49	0.43	544.14	365.87	207.33	169.21
tai100a	2.44	2.11	2.38	1.88	972.65	558.84	266.22	251.93
tai100b	0.43	0.25	0.95	0.60	712.14	647.55	334.10	158.34
tai100c	1.07	0.97	0.77	0.83	924.91	632.12	328.57	213.78
tai100d	1.77	1.67	1.66	1.72	949.81	563.30	373.92	184.99
Average(tai100)	1.43	1.25	1.44	1.26	889.88	600.45	325.70	202.26
tai150a	2.49	2.22	1.73	2.14	1990.61	1584.44	1027.63	618.60
tai150b	4.70	4.72	4.61	4.74	2746.97	1627.59	1102.50	555.33
tai150c	2.57	3.05	2.41	2.82	2096.15	1594.07	919.62	494.97
tai150d	2.30	1.94	1.93	2.07	2006.49	1366.15	966.80	615.60
Average(tai150)	3.01	2.98	2.67	2.94	2210.06	1543.06	1004.14	571.13
AVERAGE	1.64	1.55	1.53	1.54	1214.69	836.46	512.39	314.20

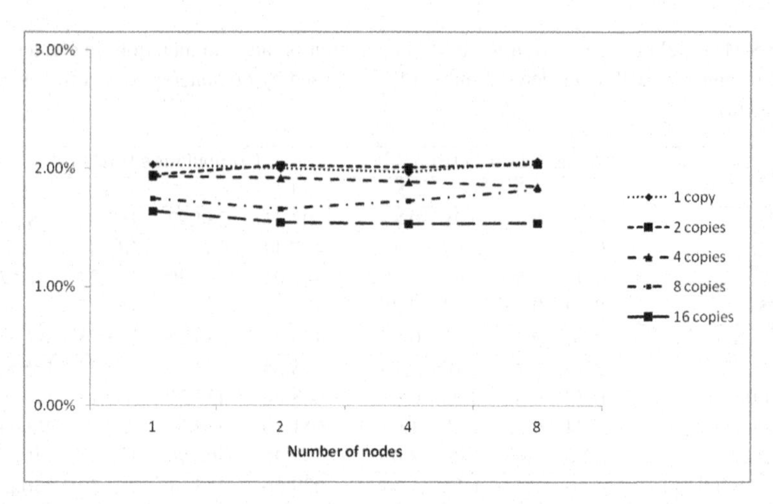

Fig. 2 Comparison of mean relative errors from the best known solution calculated for all instances, all considered number of nodes used by *Optimizing Agents*, and all cases where different number of copies of each agent were used

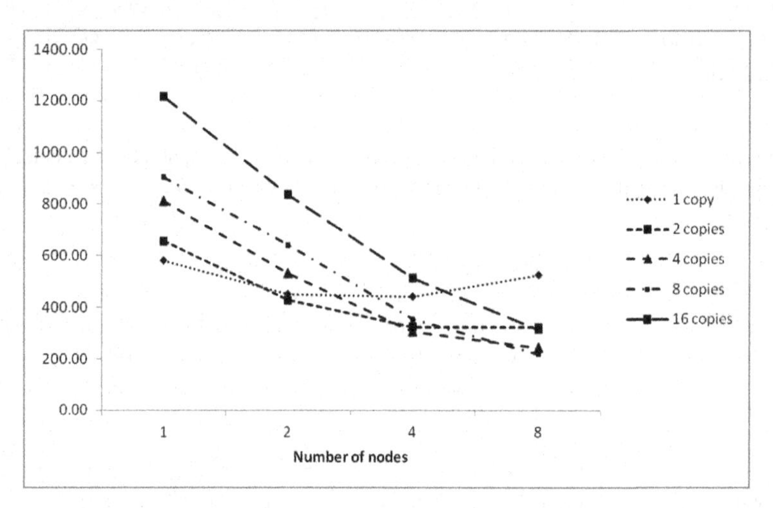

Fig. 3 Comparison of computation time calculated for all instances, all considered number of nodes used by *Optimizing Agents*, and all cases where different number of copies of each agent were used

of the observed measures have been added after each group of instances including 75, 100, and 150 customers, and the overall average to the last rows of each table. In order to improve the readability and analysis the obtained results, they are also graphically presented in Figures 2 and 3.

Analysis of the results presented in Tables 2-6 and in Figures 2 and 3 allow one to draw several interesting observations. The first general observation is that the presented approach produces good results while solving CVRP instances. Mean relative error varies from 0% to 5% for almost all instances. Definitely, the better results have been observed for instances including 75 customers than for larger ones. The differences between the obtained results for instances with 75 customers and the best known solutions very rarely exceed 1%. Also, a small mean relative error, not exceeding 2-3%, has been observed for instances, where 100 customers are to be served by a fleet of the available vehicles. Only the largest instances (150 customers) have been solved with final results worse than best known one by at most 5-6%.

By focusing observation on computation time, one can unfortunately conclude, that solving instances of the given problem sometimes requires a high computational effort measured as a time needed to reach satisfactory results. Moreover, if the problem size has increased, an increasing of the computation time has been also observed.

Results presented in the above tables and figures provide additional conclusions directly related to performance of the presented approach with a distributed computations. The smallest value of mean relative error averaged over all tested instances has been found for computations performed on the number of nodes greater than one. It seems to be true for almost all teams with different number of copies of each agent, although the difference does not seem to be significant. On the other hand, one can also conclude that if the number of copies of each agent grows, then mean relative errors decrease, regardless of the number of nodes. The teams consisting of 8 and 16 agents generally solve instances of the problem in most effective way.

Considering computation times, it can be observed that the time needed by the system while solving instances of the problem increases when the size of the team of agents is higher. On the other hand, it decreases when additional nodes are added to the main container. It is more evident in cases where the number of copies of each agent is greater than one.

In order to evaluate the level of acceleration of computation, apart from the above traditional approaches to evaluate the performances of parallel metaheuristics, also the *speedup* factor (s_m) belongs to the most commonly used. It compares the serial against the parallel time to solve a particular problem instance and is defined as $s_m = T_1/T_m$, where T_1 is execution time on one processor, and T_m is the execution time for an algorithm using m processors. Also, the *effciency* ($e_m = s_m/m$), which is a normalization of the speedup [1], [2], has been used in this context.

Unfortunately, although these measures are adequate for exact methods, they are more difficult to use when heuristic methods are engaged in search process. Such methods do not guarantee obtaining optimal solution, and they often stop before optimal solution is reached. Thus, in order to normalize obtained results for speedup and efficiency calculations, it has been decided to compare computational time needed to reach the solution with mean relative error equal to 5%, for all teams consisting of 1, 2, 4, 8 and 16 copies of each agent, respectively, and distributing over 1, 2, 4, and 8 nodes. The resulting values of speedup and efficiency factors are illustrated in Figure 4.

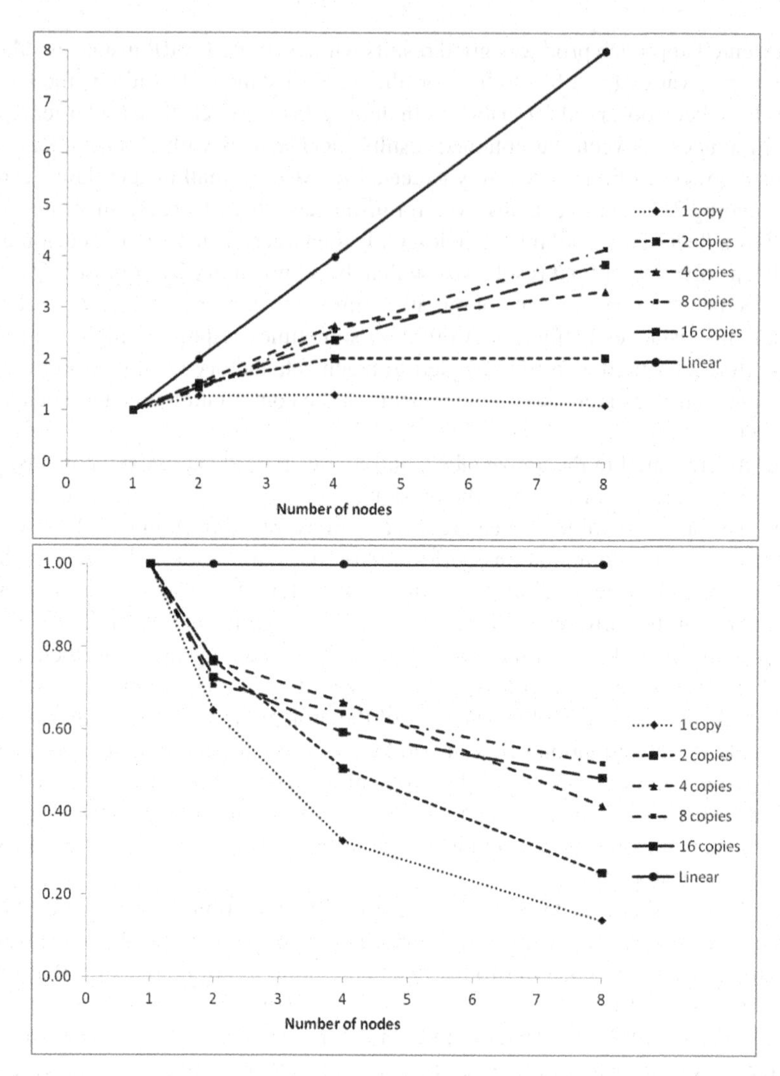

Fig. 4 Speedup (upper graph) and efficiency (lower graph) of computation time calculated for all instances, all considered number of nodes used by *Optimizing Agents*, and all cases where different number of copies of each agent were used (mean relative error equal to 5%)

Although the speedup of computation is observed for all cases with different number of agent's copies working together within a team, the recorded level of speedup is different in these cases. It is easy to see, that computation time is practically the same for the team consisting of one instance of each agent, regardless of the number of containers used for calculations. On the other hand, the significant decrease of the computation time is observed when the number of agent's copies are increased to 2, and next, to 4. Further increase of the speedup, when the number of nodes increases to 8, is observed only for teams, where 4 or 8 copies of each agent are engaged in search process. In the remaining teams, change of speedup factor is not significant.

In order to verify the above conclusions, the final part of computational experiment analysis includes the results of two non-parametric Friedman tests (for mean relative error and computation time, respectively) based on data presented in the Tables 2-6. Two hypothesis have been defined:

- H_0: performance of the proposed multi-agent system is statistically the same (in terms of mean relative error or computation time) regardless of the number of containers (nodes) used by teams of *Optimizing Agents*, while solving instances of CVRP,
- H_A: performance of the proposed multi-agent system differs (in terms of mean relative error or computation time) regardless of the number of containers (nodes) used by teams of *Optimizing Agents*, while solving instances of CVRP.

The significance level α has been set 0.05. Four treatments (different number of nodes considered in the experiment) and twelve blocks (instances) have been distinguished in the test. Four point scale, required by the test, has been used to assign weights to the results produced by the system for given instance and for each number of nodes used by system (one point has been set to the worst case, four to the best one). The value of χ^2 distribution with three degrees of freedom is equal to 7.81473. The Friedman test has been performed separately for each number of copies of each agent working within a team. The calculated values of the χ^2 statistics with four different number of nodes and twelve instances are presented in Table 7, separately for results measured as mean relative error and computation time.

It is easy to see, that for mean relative error, the test does not confirmed statistically differences between obtained results. In case of computation time, hypothesis about statistically identical performance of the system for all number of nodes considered in the experiment should be rejected for all teams with different copies of each software agent working within a team. Figures 5 and 6 present comparison of overall total of weights for both measures - mean relative error and computation time, separately.

Table 7 Results of Friedman test

Number of copies	Mean relative error		Time		χ^2 critical
of each Optimizing Agent	χ^2 statistics	Decision H_0	χ^2 statistics	Decision H_0	ststistics
1	0.7	Accept	18.7	Reject	
2	0.4	Accept	30.7	Reject	
4	1.1	Accept	33.3	Reject	7.81473
8	1.9	Accept	34.9	Reject	
16	4.5	Accept	34.9	Reject	

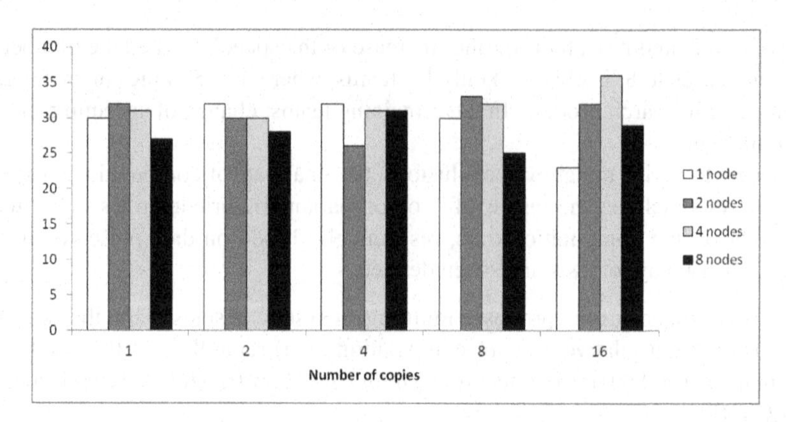

Fig. 5 Results of Friedman test: overall total of weights for each each number of copies of each agent and for each number of additional nodes used by these agents (mean relative error case)

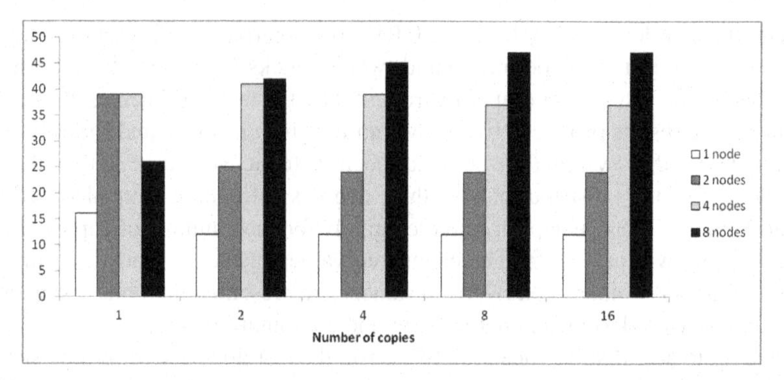

Fig. 6 Results of Friedman test: overall total of weights for each each number of copies of each agent and for each number of additional nodes used by these agents (computation time case)

5 Conclusions

Solving large scale instances of the computationally difficult optimization problems often requires using a considerable computational effort in order to reach the satisfactory results in a reasonable amount of time. The nature of metaheuristics used for solving such problems or specific features of the problems give the possibility of using parallel computation, understand as a form of computation in which different processes working simultaneously on several processors/machines solve a given problem instance. It is expected that parallel metaheuristics will outperform their sequential counterparts in terms of quality of the generated solutions as well as reducing the computation time. Another expected result in case of using parallel heuristics is their robustness, understand as an ability to solve different instances of the problem with the same effectiveness.

Last years, an agent paradigm has emerged as an interesting alternative for effectively solving different problems. Such specific features of software agents, like autonomy, reactiveness, mobility, or ability to work in teams, provide a powerful and promising tool for solving real-world problems in a range of industrial and commercial applications.

The chapter focuses on JABAT system, dedicated for (parallel/distributed) solving hard optimization problems. The main goal of the chapter was to evaluate to what extent a distribution of computation by moving *Optimizing Agents* engaged in the search process in JABAT system over available containers (nodes) can influence computational results measured by mean relative error and computation time. The experiment, carried out on selected instances of Capacitated Vehicle Routing Problem, has confirmed the existence of dependence of both observed factors on number of containers (nodes) used in experiment and the number of copies of each *Optimizing Agent*.

Among interesting direction of the planned future work of the author is concentrating on implementation of another form of parallelization of computation in JABAT. One possibility is to provide ability of moving other agents (not only *Optimizing Agents*) or the teams of agents to additional containers. The second one may focus on specific features of the problem, for example in case of vehicle routing problems, a set of customer may be initially clustered, and next, several subproblems, may be solved independently by separated search threads running in parallel.

Acknowledgements. The research has been supported by the Ministry of Science and Higher Education no. N N519 576438 for years 2010-2013. Calculations have been performed in the Academic Computer Centre TASK in Gdansk, Poland.

References

1. Alba, E., Luque, G.: Evaluation of Parallel Metaheuristics. In: Runarsson, T.P., Beyer, H.-G., Burke, E.K., Merelo-Guervs, J.J., Whitley, L.D., Yao, X. (eds.) PPSN 2006. LNCS, vol. 4193, pp. 9–14. Springer, Heidelberg (2006)
2. Barr, R.S., Hickman, B.L.: Reporting computational experiments with parallel algorithms: issues, measures, and experts opinions. ORSA Journal of Computing 5(1), 2–18 (1993)
3. Barbucha, D., Czarnowski, I., Jędrzejowicz, P., Ratajczak, E., Wierzbowska, I.: An Implementation of the JADE-base A-Team Environment. International Transactions on Systems Science and Applications 3(4), 319–328 (2008)
4. Barbucha, D., Czarnowski, I., Jędrzejowicz, P., Ratajczak, E., Wierzbowska, I.: e-JABAT - An Implementation of the Web-Based A-Team. In: Nguyen, N.T., Jain, L.C. (eds.) Intelligent Agents in the Evolution of Web and Applications. SCI, vol. 167, pp. 57–86. Springer, Heidelberg (2009)
5. Barbucha, D., Czarnowski, I., Jędrzejowicz, P., Ratajczak-Ropel, E., Wierzbowska, I.: Parallel Cooperating A-Teams. In: Jędrzejowicz, P., Nguyen, N.T., Hoang, K. (eds.) ICCCI 2011, Part II. LNCS, vol. 6923, pp. 322–331. Springer, Heidelberg (2011)

6. Barbucha, D.: Experimental Study of the Population Parameters Settings in Cooperative Multi-Agent System Solving Instances of the VRP. Submitted to LNCS Transactions on Computational Collective Intelligence (2012)
7. Bellifemine, F., Caire, G., Greenwood, D.: Developing Multi-Agent Systems with Jade. John Wiley & Sons, Chichester (2007)
8. Crainic, T.G., Nourredine, H.: Parallel meta-heuristics applications. In: Alba, E. (ed.) Parallel Metaheuristics: A New Class of Algorithms, pp. 447–494. Wiley, Hoboken (2005)
9. Crainic, T.G., Toulouse, M.: Parallel Meta-heuristics. In: Gendreau, M., Potvin, J.-Y. (eds.) Handbook of Metaheuristics. International Series in Operations Research and Management Science, vol. 146, pp. 497–541. Springer, Heidelberg (2010)
10. Doerner, K.F., Hartl, R.F., Kiechle, G., Lucká, M., Reimann, M.: Parallel Ant Systems for the Capacitated Vehicle Routing Problem. In: Gottlieb, J., Raidl, G.R. (eds.) EvoCOP 2004. LNCS, vol. 3004, pp. 72–83. Springer, Heidelberg (2004)
11. Doerner, K.F., Hartl, R.F., Benkner, S., Lucka, M.: Cooperative savings based ant colony optimization - multiple search and decomposition approaches. Parallel Processing Letters 16(3), 351–369 (2006)
12. FIPA - The Foundation for Intelligent Physical Agents, `http://www.fipa.org/` (cited May 10, 2012)
13. Garcia, B.L., Potvin, J.-Y., Rousseau, J.M.: A parallel implementation of the tabu search heuristic for vehicle routing problems with time window constraints. Computers and Operations Research 21(9), 1025–1033 (1994)
14. Gilbert, D., Aparicio, M., Atkinson, B., Brady, S., Ciccarino, J., Grosof, B., OConnor, P., Osisek, D., Pritko, S., Spagna, R., Wilson, L.: IBM Intelligent Agent Strategy. White Paper (1995)
15. Jade - Java Agent Development Framework, `http://jade.tilab.com/` (cited May 10, 2012)
16. Jennings, N.R., Sycara, K., Wooldridge, M.: A Roadmap of Agent Research and Development. Autonomous Agents and Multi-Agent Systems 1, 7–8 (1998)
17. Laporte, G., Gendreau, M., Potvin, J., Semet, F.: Classical and modern heuristics for the vehicle routing problem. International Transactions in Operational Research 7, 285–300 (2000)
18. Le Bouthillier, A., Crainic, T.G., Kropf, P.: A Guided Cooperative Search for the Vehicle Routing Problem with Time Windows. IEEE Intelligent Systems 20(4), 36–42 (2005)
19. Lin, S.: Computer solutions of the traveling salesman problem. Bell Syst. Tech. J. 44, 2245–2269 (1965)
20. Maes, P.: Artificial Life Meets Entertainment: Life like Autonomous Agents. Communications of the ACM 38(11), 108–114 (1995)
21. Meignan, D., Creput, J.C., Koukam, A.: A coalition-based metaheuristic for the vehicle routing problem. In: Proc. of the IEEE Congress of Evolutionary Computation (CEC 2008), pp. 1176–1182. IEEE Press, Hong-Kong (2008)
22. Milano, M., Roli, A.: MAGMA: a multiagent architecture for metaheuristics. IEEE Transaction on Systems, Man, and Cybernetics, Part B: Cybernetics 34(2), 925–941 (2004)
23. Osman, I.H.: Metastrategy simulated annealing and tabu search algorithms for the vehicle routing problem. Annals of Operations Research 41, 421–451 (1993)
24. Peine, H.: Application and programming experience with the area mobile agent system. Journal of Software: Practice and Experience 32(6), 515–541 (2002)
25. Rego, C.: Node ejection chains for the vehicle routing problem: sequential and parallel algorithms. Parallel Computing 27, 201–222 (2001)

26. Russell, S.J., Norvig, P.: Artificial Intelligence: a Modern Approach, 2nd edn. Prentice Hall, Upper Saddle River (2003)
27. Taillard, E.D.: Parallel iterative search methods for vehicle routing problems. Networks 23, 661–673 (1993)
28. Taillard, E.D.: VRP Instances, `http://mistic.heig-vd.ch/taillard/problemes.dir/vrp.dir/vrp.htm` (cited May 10, 2012)
29. Talbi, E.-G.: Metaheuristics: From Design to Implementation. John Wiley & Sons, Hoboken (2009)
30. Talukdar, S., Baeretzen, L., Gove, A., de Souza, P.: Asynchronous teams: Cooperation schemes for autonomous agents. Journal of Heuristics 4, 295–321 (1998)
31. White, J.E.: Telescript technology: The foundation for the electronic marketplace. White paper. General Magic, Inc. (1994)
32. Wooldridge, M.J., Jennings, N.R.: Intelligent Agents: Theory and Practice. Knowledge Engineering Review 10(2), 115–152 (1995)

Structure vs. Efficiency of the Cross-Entropy Based Population Learning Algorithm for Discrete-Continuous Scheduling with Continuous Resource Discretisation

Piotr Jędrzejowicz and Aleksander Skakovski*

Abstract. In the chapter, we consider the population learning algorithm (PLA2), earlier designed by the authors, and study how the interconnection topology and heterogeneity of the constituent modules influence its efficiency. PLA2 is a population-based approach which takes advantage of the features common to the social education system rather than to the evolutionary processes. The problem of scheduling nonpreemtable tasks on parallel identical machines under constraint on discrete resource and requiring, additionally, renewable continuous resource to minimize the schedule length is chosen as the problem to cope with. A continuous resource is divisible continuously and is allocated to tasks from given intervals in amounts unknown in advance. Task processing rate depends on the allocated amount of the continuous resource. To eliminate time consuming optimal continuous resource allocation, an NP-hard problem Θ_Z with continuous resource discretisation is introduced and sub-optimally solved by PLA2. The PLA2's island design can be easily transferred to an agent system with cooperating agents.

1 Introduction

A problem of scheduling jobs on multiple machines under constraint on discrete resource and requiring, additionally, renewable continuous resource to minimize the schedule length is considered in the chapter. In the problem two types of resources are considered: discrete and continuous. A discrete resource is divisible

Piotr Jędrzejowicz · Aleksander Skakovski
Department of Information Systems, Gdynia Maritime University,
Morska 83, 81-225 Gdynia, Poland
e-mail: {pj,askakow}@am.gdynia.pl

* Corresponding author.

I. Czarnowski et al. (Eds.): Agent-Based Optimization, SCI 456, pp. 77–102.
DOI: 10.1007/978-3-642-34097-0_4

discretely, for example a set of machines or a set of mechanical or pumping machines. A continuous resource is divisible continuously and is allocated to the jobs from given intervals in amounts unknown in advance. In practice a continuous resource may be limited in amount - for example power (electric, pneumatic, hydraulic) supplying a set of machines, limited gas flow intensity supplying forge furnaces in a steel plant, or limited fuel flow intensity in refueling terminals.

The problem of scheduling jobs on multiple machines under constraint on discrete resource and requiring, additionally, renewable continuous resource was intensively explored in [9], [10], [11], [12], and we define the problem in the same way. Namely, we consider n independent, nonpreemptable jobs, each of them simultaneously requiring for its processing at time t a machine from a set of m parallel, identical machines (the discrete resource) and an amount (unknown in advance) $u_i(t) \in [0, 1]$, $i = 1, 2, \ldots, n$, of a continuous renewable resource. The job model is given in the form:

$$\dot{x}_i(t) = \frac{dx_i(t)}{d(t)} = f_i[u_i(t)],\ x_i(0) = 0,\ x_i(C_i) = \tilde{x}_i\ , \tag{1}$$

where $x_i(t)$ is the state of job i at time t, f_i is an increasing continuous function, $f_i(0) = 0$, C_i is (unknown in advance) completion time of job i, and $\tilde{x}_i$ is its processing demand (final state). We assume, without loss of generality, that $\sum_{i=1}^{n} u_i(t) = 1$ for every t. The problem is to find a sequence of jobs on machines and, simultaneously, a continuous resource allocation that minimizes the given scheduling criterion. The problem is computationally complex and is at least as hard as the classical RCPSP (Resource Constrained Project Scheduling Problem), since the existence of an additional continuous resource cannot make the problem any simpler [11], [12]. The defined problem can be decomposed into two interrelated sub problems: (i) to find a feasible sequence of jobs on machines, and (ii) to allocate the continuous resource among jobs already sequenced. The notion of a feasible sequence is of crucial importance. According to [10] a feasible schedule can be divided into $p \leq n$ intervals defined by completion times of consecutive jobs. Let Z_k denote the combination of jobs processed in parallel in the k-th interval. Thus, in general, a feasible sequence FS of combinations Z_k, $k = 1, 2, \ldots, p$, can be associated with each feasible schedule. Feasibility of such a sequence requires that the number of elements in each combination does not exceed m and that each job appears exactly in one or in consecutive combinations in FS (nonpreemptability). It has been shown in [9] that for concave job models and the schedule length minimization problem, it is sufficient to consider feasible sequences of combinations Z_k, $k = 1, 2, \ldots, n - m + 1$, composed of exactly m jobs each. For a given feasible sequence FS of jobs on machines, we can find an optimal continuous resource allocation, i.e. an allocation that leads to a schedule minimizing the given criterion from among all feasible schedules generated by FS. At this point, a convex mathematical programming problem has to be solved, in the general case (see [9]). An optimal schedule for a given feasible sequence (i.e. a schedule resulting from an optimal continuous resource allocation for this sequence) is called a semi-optimal schedule. In consequence, a globally optimal schedule can be found by solving the continuous resource allocation problem optimally for all feasible sequences. Unfortunately, in general, the number of feasible sequences grows exponentially with the number of jobs. Therefore it is justified to apply some approximation algorithm or metaheuristic.

Because finding an optimal allocation of a continuous resource to a feasible schedule requires using specialized and time-consuming solver, an idea of continuous resource discretisation was proposed in [12]. We use the same approach in the chapter. Namely, we assume that the number of possible continuous resource allocations to a task J_i is D_i, i.e. is fixed, and the amount of the continuous resource for each $l_i = 1, 2, \dots, D_i$ is known in advance (in the original problem there was infinite number of the continuous resource allocations to a task and the amount of the continuous resource to be allocated was not known in advance). Because a different amount of the continuous resource is allocated to task J_i for each l_i, l_i is called a processing mode of task J_i. Such discretisation of the continuous resource allows treating it as a discrete resource.

The problem of scheduling jobs on multiple machines under additional continuous resource with continuous resource discretisation is NP-hard [12]. A population-learning algorithm (SLA) first proposed in [6] was used to tackle the problem, since it was effective in solving other scheduling problems considered in [5], [3], [4]. Promising results obtained by the proposed in [8] version of PLA - PLA1 proved the approach for solving Θ_Z to be effective and caused the design of PLA2 proposed in [8]. PLA2 uses four main procedures: a cross-entropy (CE), a Tabu Search (TS) procedure, an island-based evolutionary algorithm (IBEA), and a population-based evolutionary algorithm (PBEA). All mentioned procedures could be viewed as independent and cooperating agents and used to design an agent system. Because all the procedures used in PLA2 were thoroughly described in [7] and [8] we only briefly remind the procedures in this work in Sections 3.1-3.4 respectively.

The main goal of our research was to find out whether the interconnection topology of a learning stages (or islands), might have some effect on the algorithm's efficiency. For this reason we proposed six versions of PLA2 that differ from each other by their structure and migration scheme. We assume that the efficiency of the algorithm is its ability to yield "good" quality solutions of a problem within a given number of fitness function evaluations. On this basis we have compared all proposed versions of PLA2 making them to carry out the same or approximately the same number of fitness function evaluations. Assuming such approach it is easier to judge on the efficiency of the proposed versions of PLA2 by only comparing the quality of the solutions they yielded. A computational experiment, described in Section 4, was carried out to test the influence of the interconnection topology of the available islands and the possible influence of migration size between CE-island and IBEA-islands on the quality of the PLA2 found solutions.

2 Problem Formulation

We define a problem Θ_Z in the same way as in [12]. Namely, let $\boldsymbol{J} = \{J_1, J_2, \dots, J_n\}$ be a set of nonpreemtable tasks, with no precedence relations and ready times $r_i = 0$, $i = 1, 2, \dots, n$, and $\boldsymbol{P} = \{P_1, P_2, \dots, P_m\}$ be a set of parallel and identical machines, and there is one additional renewable discrete resource in amount $U = 1$ available. A task J_i can be processed in one of the modes $l_i = 1, 2, \dots, D_i$ (D_i – the number of processing modes of task J_i), for which J_i requires a machine from $\boldsymbol{P}$ and amount of the additional resource known in advance. The processing mode of J_i cannot change during the processing. For each task two vectors are defined: a processing times vector $\tau_i = [\tau_i^1, \tau_i^2, ..., \tau_i^{D_i}]$, where $\tau_i^{l_i}$ is the processing time of task J_i in mode $l_i = 1, 2, \dots, D_i$ and a vector of additional resource quantities allocated in each processing mode $u_i = [u_i^1, u_i^2, ..., u_i^{D_i}]$. The problem is to find processing modes for tasks from $\boldsymbol{J}$ and their sequence on machines from $\boldsymbol{P}$ such that schedule length $Q = \max\{C_i\}$, $i = 1, \dots, n$ is minimized.

3 Population Learning Algorithm

Population learning algorithm proposed in [6] has been inspired by analogies to a social phenomenon rather than to evolutionary processes. The population learning algorithm takes advantage of features that are common to social education systems:

- A generation of individuals enters the system.
- Individuals learn through organized tuition, interaction, self-study and self-improvement.
- Learning process is inherently parallel with different schools, curricula, teachers, etc.
- Learning process is divided into stages.
- More advanced and more demanding stages are entered by a diminishing number of individuals from the initial population (generation).
- At higher stages more advanced education techniques are used.
- The final stage can be reached by only a fraction of the initial population.

All individuals (solutions) used in the PLA2 procedure can be characterized in the following manner:

- an individual (a solution) is represented by an n-element vector $S = [c_i | \ 1 \le i \le n]$,
- all processing modes of all tasks are numbered consecutively. Thus processing mode l_b of task J_b has the number $c_b = \sum_{i=1}^{b-1} D_i + l_b$,
- all S representing feasible solutions are potential individuals,

- each individual can be transformed into a schedule by applying LSG, which is a specially designed list-scheduling algorithm for discrete-continuous scheduling,
- each schedule produced by the LSG can be directly evaluated in terms of its fitness.

The PLA2 model can be also viewed as an island model, were islands are connected to each other according to some topology and exchange individuals in order to collectively find best possible solution to the problem. Such island-based design can be easily transferred to an agent system with cooperating agents. We used three kinds of learning procedures to design PLA2: cross-entropy (CE), Tabu search (TS), and an island-based evolutionary algorithm (IBEA) all combined into some structures. As a learning procedure IBEA uses population-based evolutionary algorithm (PBEA) to evolve a population on an island, which is described in Section 3.3. PBEA is also used to evolve solutions on an island independently on IBEA, for example in case with random solution migration among islands. We distinguish two categories of island groups – heterogeneous and homogeneous, dependently on the type of the learning procedures carried out on the islands. We refer to the group of islands as heterogeneous, if the learning procedures carried out on at least one island is different from the learning procedures carried out on the rest of the islands in the group. We refer to the group of islands as homogeneous, if the same learning procedure is carried out on each island in the group. In our work, we will refer to a particular island as heterogeneous (Ht), if CE or TS procedure is carried out on it, and homogeneous (Hm), if PBEA is carried out on it. We will use the terms a learning procedure and an island interchangeably. The main goal of our research was to find out whether a topology of a learning stages (or islands), might have some effect on the algorithm's efficiency. For this reason we proposed several versions of PLA2 that differ from each other by their structure and migration scheme. We will refer to these versions of PLA2 as algorithms, and some letter code will be assigned to each of them. In order to distinguish the algorithms, we considered two their basic types, each of them having two topology schemes. In the first basic type, all islands participate in the solution evolution and migration at least once, but only *selected* islands take part in the cyclic solution migration among islands (the letter code for this version will contain letter "S"). In the second basic type – all islands take part in the cyclic solution migration among islands (the letter code for this version will contain letter "A"). As it was mentioned above, each basic type appears in two topology schemes. In the first topology scheme, islands are located on a directed ring and the individuals migrate among the islands along the ring (the letter code for this scheme will contain letter "O" and we will refer in the following text to this topology as a *ring topology*). In the second topology scheme – individuals migrate between randomly chosen pairs of the islands (the letter code for this scheme will contain letter "X" and we will refer in the following text to this topology as a *random topology*). Moreover, the letter code for the algorithms in which CE procedure sends multiple solutions to the island-in-pair during the migration phase will contain letter "m". For the version where CE procedure sends a single solution to the island-in-pair during the migration phase the letter code will contain letter “s”. Therefore, the letter code “AO-m” stands for the algorithm in which all islands comprise a directed ring of heterogeneous islands and procedure CE sends multiple solutions to the island-in-pair during the migration phase. In our present research, we consider six versions of PLA2, namely: SO, SX, AO-m, AO-s, AX-m, and AX-s. Because all proposed algorithms are versions of PLA2, they have common phases, which are shown in a generalized versions as S- or A-algorithms. The pseudo codes, as well as figures illustrating all the proposed algorithms are given below. In a simplified graphic illustration of the algorithms in Figures 1 - 4, solid lines show islands participating in the cyclic solution migration and dash-dot lines show islands where learning procedures are carried out only once.

S-algorithm
Begin
  Create an initial population P_0 of the size x_0 - 1 using procedure cross-entropy (CE).
  Create an individual *TSI* in which all tasks J_i are to be executed in mode l_i = 1 (a mode characterized by minimal quantity of additional resource u_i^1 and maximal task processing time τ_i^1, $1 \le i \le n$).
  Improve the individual *TSI* with the tabu search (TS) procedure.
  Create population $P_1 = P_0 + TSI$.
  Distribute equally individuals from P_1 among all Hm-islands.
  Carry out the appropriate Learning stage SO or SX designed for SO and SX algorithms respectively.
  Output the best solution to the problem.
End.

Learning stage SO
Begin
  Improve individuals on Hm-islands with procedure IBEA.
End.

Learning stage SX
Begin
  Improve individuals on each Hm-island with procedure PBEA, cyclically exchanging best solutions between randomly chosen pairs of Hm-islands.
End.

In all proposed algorithms, $x_0 = K \cdot PS$, where *K* – the number of homogeneous islands and *PS* – the population size on an island defined in procedure IBEA.

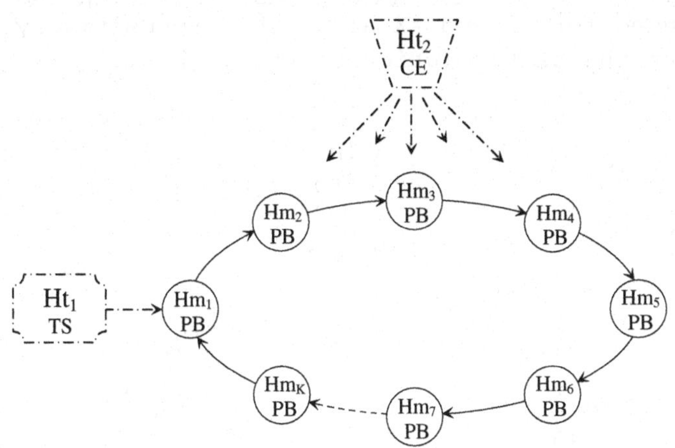

Fig. 1 A simplified scheme of SO algorithm

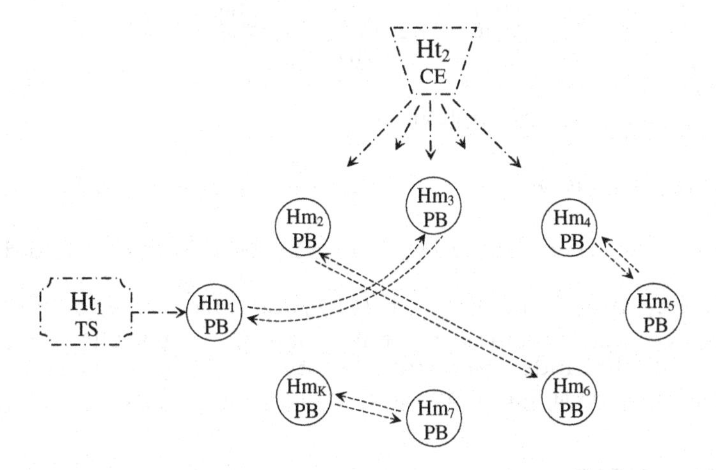

Fig. 2 A simplified scheme of SX algorithm

A-algorithm

Begin

Create an initial population P_0 of the size x_0 using cross-entropy procedure (CE).

Distribute equally individuals from P_0 among all Hm-islands.

Create an individual *TSI* in which all tasks J_i are to be executed in mode $l_i = 1$ (a mode characterized by minimal quantity of additional resource u_i^1 and maximal task processing time τ_i^1, $1 \le i \le n$).

Send *TSI* to the Tabu Search (TS) procedure.

Carry out the appropriate Learning stage AO or AX designed for AO and AX algorithms respectively.
Output the best solution to the problem.
End.

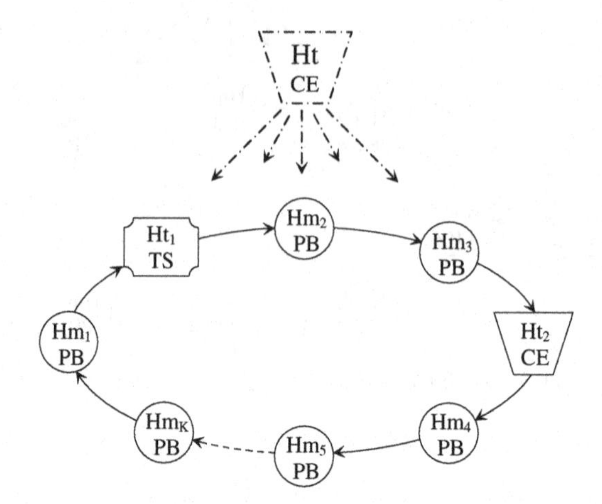

Fig. 3 A simplified scheme of AO algorithm

Learning stage AO
Begin
Create a directed ring of all available islands as follows:
Hm_1, Ht_1(TS), Hm_2, Hm_3, Ht_2(CE), … , Hm_K, where K • 3.
Improve individuals on the islands with the assigned to the islands procedures cyclically sending best solution from each island along the ring.
End.

Learning stage AX
Begin
Improve individuals on all available islands with the assigned to the islands procedures cyclically exchanging best solution between randomly chosen pairs of islands.
End.

In the AO algorithm, CE procedure receives multiple solutions from the homogeneous islands Hm_1, Hm_2 and Hm_3, and sends a single solution to Hm_4 (or Hm_1, when $K = 3$) in AO-s algorithm, or multiple solutions in AO-m algorithm. In AX algorithm CE procedure also receives multiple solutions from Hm_1, Hm_2 and Hm_3, and sends to the randomly chosen island a single solution in AX-s algorithm, or multiple solutions in AX-m algorithm. On all homogeneous islands Hm_i, $i = 1, 2, \ldots, K$ we used PBEA procedure.

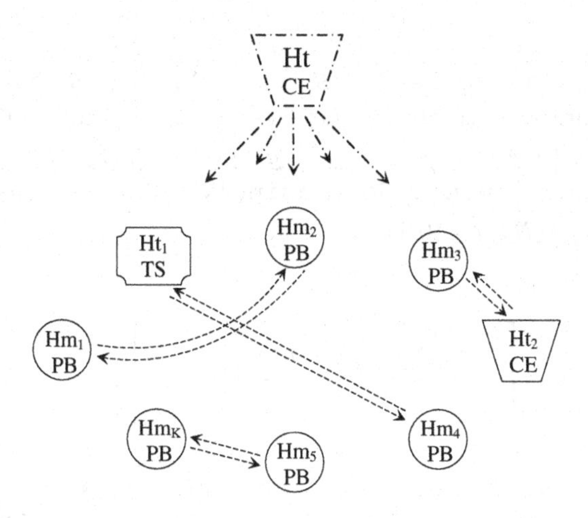

Fig. 4 A simplified scheme of AX algorithm

3.1 A Cross-Entropy Algorithm

In PLA2 the proposed CE procedure is perceived as the procedure preparing some solution basis for further improvement by procedure IBEA. In CE procedure a cross-entropy (CE) method first proposed in [13] is used since it was effective in solving various difficult combinatorial optimization problems [1]. It follows from the definition of the solution vector *S* that a number c_i in S unequivocally identifies a task and the task processing mode. In order to use CE method, we would like to know the probability of locating a task J_i on a particular place *j* in the vector. For this reason we introduce two success probability vectors $\hat{p}_j$ and $\hat{p}'_{ji}$ related to each task J_i and its place *j* in solution *S*. Vector $\hat{p}_j = [p_{ji} \mid 1 \le i \le n]$, $1 \le j \le n$ contains p_{ji} values, which is the probability that on the place *j* there will be located a task *i*. Vector $\hat{p}'_{ji} = [p_{jil} \mid 1 \le l \le D_i]$, $1 \le j \le n$, $1 \le i \le n$ contains p_{jil} values, which is the probability that on place *j* task *i* will be executed in mode *l*. A procedure CE using cross-entropy method for combinatorial optimization described in [1] and modified for solving Θ_Z problem is shown in the following pseudo code:

```
Procedure CE
Begin
 Set ic = 1 (ic - iteration counter), ic^stop - maximal
 number of iterations, a:= 1.
```

Set $\hat{p}_j = [p_{ji} = 1/n \mid 1 \le i \le n], 1 \le j \le n.$

Set $\hat{p}'_{ji} = [p_{jil} = 1/D_i \mid 1 \le l \le D_i], 1 \le j \le n, 1 \le i \le n.$

While $ic \le ic^{stop}$ do

Generate a sample S_1, S_2, ... , S_s, ... , S_N of solutions with success probability vectors $\hat{p}_j$ and $\hat{p}'_{ji}$.

Order S_1, S_2, ... , S_s, ... , S_N by values of their fitness function non-decreasingly.

Set $\gamma = \lceil \rho \cdot N \rceil, \rho \in (0,1)$.

Set

$$\hat{p}_j = \left[p_{ji} = \frac{\sum_{s=1}^{\gamma} I(S_s(j) = i)}{\gamma} \middle| 1 \le i \le n \right], \tag{2}$$

$1 \le j \le n$, $I(S_s(j) = i) = 1$, $I(S_s(j)$ • $i) = 0$, where $S_s(j)$ - number of the task located on j-th place in s-th solution S.

Set

$$\hat{p}'_{ji} = \left[p_{jil} = \frac{\sum_{s=1}^{\gamma} I(S_s(ji) = l)}{\gamma} \middle| 1 \le l \le D_i \right], \tag{3}$$

$1 \le j \le n$, $1 \le i \le n$, $I(S_s(ji) = l) = 1$, $I(S_s(ji) \ne l) = 0$, where $S_s(ji)$ - an execution mode of task i located on j-th place in s-th solution S.

Save the first $h = \lceil K \cdot PS / ic^{stop} \rceil$ best solutions from the ordered sample into P_0 under address a. Set $a := a + h$.

Set $ic := ic + 1$.

EndWhile.

EndProcedure.

In the presented pseudo code, a parameter N is the number of solutions in a sample generated in each iteration. A parameter ρ determines the percentage of the best solutions in the current sample that are used to calculate new values for the vectors $\hat{p}_j$ and $\hat{p}'_{ji}$. The both parameters were determined empirically and set $N = 1000$ and $\rho = 0{,}2$. Parameters K – the number of islands and PS – the population size are defined in procedure IBEA and PBEA respectively.

3.2 Tabu Search

Tabu search is another metaheuristic used in the considered versions of PLA (see [4]). To present general idea of the present implementation of the tabu search procedure we introduce the neighborhoods N_t and N_{md} of a solution S. N_t is a set of solutions generated from S by moving a task $J_i \in S$ from place i to the rest $n - 1$ places. Thus we yield $|N_t| = n \cdot (n - 1)$ neighbors. N_{md} is a set of solutions generated from S by assigning to task $J_i \in S$ one by one in a row all of its D modes, assuming that all tasks can be executed in D modes. Thus we yield another $|N_{md}| = n \cdot (D - 1)$ neighbors. The considered tabu search procedure is shown in the following pseudo code:

```
Procedure TS
Begin
 Set S0 = initial solution TSI (li = 1, 1 • i • n).
 Set the best solution Sbest = S0.
 Set Tabu List TL = Ø.
 Set Nt = {S0} and Nmd = Ø.
 Set nit = 7 (determined empirically).
 Repeat the following max_number_of_iterations times:
   Find the best legal neighbour Sbln of S0, i.e. the
   best across Nt and Nmd neighbour which is not on TL.
   Set S0 = Sbln.
   If Sbln is more fit than Sbest then Sbest = Sbln.
   Put Sbln on the Tabu list.
   If the fitness of S0 has not improved after nit num-
   ber of iterations construct a new solution by mov-
   ing a task Ji in S0 to one of the chosen randomly
   less frequently visited places on the task list and
   assigning to it one of the chosen randomly less
   frequently assigned execution modes.
 EndRepeat.
End.
```

The size of the Tabu List (TL) was determined empirically and set to 500 solutions.

3.3 An Island-Based Evolutionary Algorithm

The following pseudo-code shows main stages of the IBEA algorithm:

```
Procedure IBEA
Begin
 Set the number of islands K, the number of popula-
 tions PN to be evolved on each island.
 While no stopping criteria is met do
   For each island Ik do
      Evolve PN generations using procedure PBEA.
```

```
      Send the best solution to I_((k mod K) + 1).
      Incorporate the best solution from
      I_(((K+k -2) mod K) + 1) instead of the best one.
    EndFor
  EndWhile
  Find the best solution across all islands and save it
  as the final one.
End.
```

3.4 A Population-Based Evolutionary Algorithm

Population-based evolutionary algorithm (PBEA) proposed in [7] as a part of IBEA for solving discrete-continuous scheduling problem is used as a learning procedure to evolve solutions on homogeneous islands in all considered versions of PLA2. PBEA algorithm is shown in the following pseudo-code:

```
Procedure PBEA
Begin
  Set population size PS.
  Set ic:= 0; (ic - iteration counter).
  While no stopping criteria is met do
    Set ic:= ic + 1,
    Calculate fitness factor for each individual in
    PP_(ic-1) using LSG,
    Form new population PP_ic:
      Select randomly a quarter of PS of individuals
      from PP_(ic-1) (probability of selection depends on
      fitness of an individual).
      Produce a quarter of PS of individuals by apply-
      ing crossover operator to previously selected in-
      dividuals from PP_(ic-1).
      Produce a quarter of PS of individuals by apply-
      ing mutation operators to previously selected in-
      dividuals from PP_(ic-1).
      Generate half of a quarter of PS of individuals
      from set of potential individuals (random task
      processing mode and task order).
      Generate half of a quarter of PS of individuals
      from set of potential individuals (random task
      processing mode and ascending order of the task
      numbers).
    EndWhile
End.
```

LSG algorithm used within PBEA is carried out in three steps as follows:

```
Procedure LSG
Begin
  Construct a list of tasks from the code representing
  individuals. Set loop over tasks on the list.
  Within the loop, allocate current task to a machine
  considering the amount of a continuous resource al-
  lotted to the task, and minimizing the beginning time
  of its processing. Continue with tasks until all have
  been allocated.
  Calculate the fitness of the individual S as
  Q_u = max{C_i}, i = 1, ... , n.
End.
```

4 Computational Experiments

The proposed six versions of the cross-entropy based population learning algorithm for solving discrete-continuous scheduling problems with continuous resource discretisation were implemented and tested. The efficiencies of all six algorithms were compared to each other, as well as to the tabu search (TS) procedure, used within each of the algorithms which was run in addition as an independent algorithm. In the procedure CE, as it was mentioned earlier, parameters ρ and N were determined empirically and set $N = 1000$ and $\rho = 0{,}2$. The size of the Tabu List (TL) was determined empirically as well, and set to 500 solutions. For testing purposes three combinations of n x m were considered (n – the number of tasks and m – the number of machines): 10x2, 10x3, and 20x2. For each combination n x m 100 instances of a problem Θ_Z were generated and three discretisation levels D were considered: 10, 20, and 50. This way we considered nine sizes of the problem: 10x2x10, 10x2x20, 10x2x50, 10x3x10, … , 20x2x50, which makes 900 instances of the problem in total. In all problem instances, we have used the same as in [12] the task processing rate function calculated according to the formula:

$$f_i^{l_i} = u_i^{l_i 1/\alpha_i}, \alpha_i \in \{1, 2\}, \tag{4}$$

where α_i could take the values 1 and 2 with the same probability. The values of the task processing times were calculated according to the formula:

$$\tau_i^{l_i} = \frac{\tilde{x}_i}{f_i^{l_i}}, \; i = 1, 2, \ldots, n. \tag{5}$$

The continuous resource was discretised uniformly according to the formula:

$$u_i^{l_i} = \frac{l_i}{D_i}, \; D_i = D, \; D \in \{10, 20, 50\}, \; i = 1, 2, \ldots, n. \tag{6}$$

Each of the considered algorithms carried out about 720000 fitness function evaluations to yield one solution for the instance of the problem. Each instance was tested 43 times by all the proposed algorithms. Mean time required by the considered algorithms to find a solution for the problem sizes 10x2 and 10x3 for all discretisation levels on Pentium (R) 4 CPU 3.00GHz compiled with aid of Borland Delphi Personal v.7.0 was approximately 4 - 7s, and for the problem size 20x2 for all discretisation levels approximately 8 – 13s.

In order to evaluate the efficiency of the proposed algorithms we used such parameters as relative errors (minimum, average, maximum) of the solutions yielded by the algorithms, as well as percentage of the best found solutions of the same quality as the best-known solutions. Relative errors (RE) of the solutions compared to the best-known solutions were calculated according to the formulae $RE = (Q_{PLA2} - Q_{best\text{-}known})/Q_{best\text{-}known}$, where Q – the quality of a considered solution. The set of the best-known solutions was determined by the authors while using all designed by them procedures and algorithms, namely PBEA, IBEA, TS, PLA1, PLA2, AX-m, AX-s, SX, SO, AO-m, AO-s, for solving problem Θ_Z. We have determined RE_{min} and RE_{max} for every size of the considered problem as a minimum or respectively maximum RE across 4300 REs calculated while solving each of the 100 instances 43 times. We have also determined RE_{avg} as a mean value of 4300 REs obtained within 43 runs of 100 instances of the considered problem. The values of RE_{min}, RE_{avg} and RE_{max} of the solutions found by all proposed algorithms for all problem sizes are presented in Tables 1 - 9. The values of REs in Tables 1 - 9 show how much schedules yielded by the proposed algorithms were longer than the best known schedule for the same case. For example, in Table 1 for the case 10x2x10 for SO algorithm, RE_{avg} = 3.28% means that the schedule length of all schedules yielded by SO algorithm was on average 3.28% longer than the best-known. For the same case, RE_{max} = 9.76% means that the longest schedule among all schedules yielded by SO algorithm was 9.76% longer than the best-known. To make it easier to evaluate their efficiency, in Tables 1 - 9 below, we have ordered the algorithms according to their REs non-decreasingly.

Table 1 The algorithms ordered non-decreasingly according to their RE_{min}, RE_{avg} and RE_{max} for the size 10x2x10 of the problem Θ_z

Nr	Algm	RE_{min}	Algm	RE_{avg}	Algm	RE_{max}
1	SX	0,01%	SO	3,28%	SO	9,76%
2	AO-m	0,01%	SX	3,31%	TS	9,91%
3	AO-s	0,01%	TS	3,49%	SX	10,00%
4	AX-m	0,01%	AX-m	3,54%	AX-s	11,39%
5	AX-s	0,01%	AX-s	3,58%	AX-m	12,33%
6	TS	0,01%	AO-m	6,30%	AO-m	14,68%
7	SO	0,19%	AO-s	6,30%	AO-s	16,53%

For the problem size 10x2x10, according to the values of the RE_{min} in Table 1, SO algorithm has the largest RE_{min}, however quite close to the REs of the other algorithms. On the other hand, considering RE_{avg}, SO algorithm has the lowest RE_{avg}, and this way, is the leader in a group of the algorithms: SO, SX, TS, AX-m, AX-s, with similar RE_{avg} values. The algorithms AO-m, AO-s make another group of the same RE_{avg} values, where RE_{avg} is about twice higher than in the first group. Considering RE_{max}, it is also possible to classify the algorithms into two groups: with low RE_{max}: SO, TS, SX, and with high RE_{max}: AX-s, AX-m, AO-m, AO-s. In the first group the algorithms differ from 9,76% to 10,00%, while in the second – from 11,39% to 16,53%. For the problem size 10x2x10, $RE_{min} \in [0{,}01\%, 0{,}19\%]$, $RE_{avg} \in [3{,}28\%, 6{,}30\%]$, $RE_{max} \in [9{,}76\%, 16{,}53\%]$. Generally, according to Table 1, the algorithms exploiting the directed ring migration scheme (the ring topology) or random migration scheme (the random topology), built exclusively on homogeneous islands, as well as scheme built on all islands with the random topology perform better, than the algorithms exploiting the ring topology built on all islands - both homogeneous and heterogeneous. This way, for the size 10x2x10 – SO, SX and AX-s perform better than the rest of the algorithms.

Table 2 The algorithms ordered non-decreasingly according to their RE_{min}, RE_{avg} and RE_{max} for the size 10x3x10 of the problem Θ_z

Nr	Algm	RE_{min}	Algm	RE_{avg}	Algm	RE_{max}
1	AX-s	0,00%	SO	4,67%	SO	15,92%
2	SO	0,00%	AX-m	4,68%	AX-m	16,07%
3	AX-m	0,00%	SX	4,69%	TS	17,72%
4	SX	0,01%	AX-s	4,74%	SX	18,39%
5	TS	0,07%	TS	5,36%	AX-s	19,33%
6	AO-m	0,10%	AO-m	8,66%	AO-s	25,67%
7	AO-s	0,28%	AO-s	8,71%	AO-m	26,06%

For the problem size 10x3x10, according to the values of the RE_{min} in Table 2, algorithms AX-s, SO, AX-m were able to find the best-known solutions, and algorithms: SX, TS, AO-m, AO-s could not. However, RE_{min} values of the latter group do not differ significantly from the best-known solutions, namely, from 0,01% to 0,28%. Considering RE_{avg}, SO algorithm has the lowest RE_{avg}, and this way, is the leader in a group of the algorithms: SO, AX-m, SX, AX-s with RE_{avg} values differing from 4,67% to 4,74%. TS algorithm is in-between the first group and the third, made of AO-m and AO-s algorithms, whose RE_{avg} values are considerably higher than in the first group, i.e. 8,66% and 8,71% respectively. Speaking about RE_{max}, it is also possible to classify the algorithms into two groups of similar values of RE_{max}: SO, AX-m, TS, SX, from 15,92% to 18,39%, and a group of high RE_{max}: AX-s, AO-s, AO-m, from 19,33% to 26,06%. For the problem size 10x3x10, we also give the intervals to which belong the values of the considered parameters, i.e. $RE_{min} \in [0{,}00\%, 0{,}28\%]$, $RE_{avg} \in [4{,}67\%, 8{,}71\%]$, and finally,

$RE_{max} \in [15,92\%, 26,06\%]$. As it could be seen, the RE_{avg} and RE_{max} of all algorithms have increased while scheduling 10 tasks on 3 machines compared to scheduling 10 tasks on 2 machines. In this case again, homogeneous ring topology, as well as random topology for both homogeneous and heterogeneous structures preformed better than other algorithms. In addition, it can be noticed, that the algorithms where CE procedure sends multiple solutions during the migration phase perform better, than when it sends a single solution. To finalize, for the size 10x3x10 – SO, AX-m and SX perform better, than the other algorithms.

For the problem size 20x2x10, according to the values of the RE_{min} in Table 3, only algorithm AX-m was able to find the best-known solutions, and the rest of the algorithms - could not. According to RE_{min}, the algorithms: AX-s, SX, SO, TS make a middle group with the values from 0,26% to 0,68%. The algorithms AO-m and AO-s make the third group with RE_{min} values from 1,12% to 1,16%, and are nearly twice as high as in the middle group. Considering RE_{avg}, AX-m algorithm has the lowest RE_{avg}, and is the leader in a group of the algorithms: AX-m, AX-s, SO, SX, TS with RE_{avg} values differing from 4,76% to 6,26%. The algorithms AO-s, AO-m make the third group of the similar RE_{avg} values, where RE_{avg} is from 9,19% to 9,31% which are considerably higher than in the middle group. Considering RE_{max}, it is also possible to classify the algorithms into two groups: with low RE_{max}: SO, SX, AX-m, TS, AX-s, with the values from 11,47% to 12,54%, and another group: AO-m, AO-s, with high RE_{max} values from 16,71% to 18,19%. For the problem size 20x2x10, the intervals of the REs are as follow: $RE_{min} \in [0,00\%, 1,16\%]$, $RE_{avg} \in [4,76\%, 9,31\%]$, $RE_{max} \in [11,47\%, 18,19\%]$. For this problem size, our observations on the topology point at the random topology of both heterogeneous and homogeneous structures, as well as homogeneous ring topology as most efficient ones. Here, CE procedure sending multiple solutions during the migration phase, perform better than when it sends only a single solution. For the size 20x2x10, AX-m, SO and AX-s are the most efficient algorithms.

Table 3 The algorithms ordered non-decreasingly according to their RE_{min}, RE_{avg} and RE_{max} the size 20x2x10 of the problem Θ_z

Nr	Algm	RE_{min}	Algm	RE_{avg}	Algm	RE_{max}
1	AX-m	0,00%	AX-m	4,76%	SO	11,47%
2	AX-s	0,26%	AX-s	4,86%	SX	11,74%
3	SX	0,40%	SO	5,46%	AX-m	11,81%
4	SO	0,51%	SX	5,56%	TS	12,09%
5	TS	0,68%	TS	6,26%	AX-s	12,54%
6	AO-m	1,12%	AO-s	9,19%	AO-m	16,71%
7	AO-s	1,16%	AO-m	9,31%	AO-s	18,19%

Table 4 The algorithms ordered non-decreasingly according to their RE_{min}, RE_{avg} and RE_{max} for the size 10x2x20 of the problem Θ_z

Nr	Algm	RE_{min}	Algm	RE_{avg}	Algm	RE_{max}
1	SX	0,00%	SO	2,05%	SO	6,67%
2	AO-m	0,00%	SX	2,07%	SX	7,04%
3	AO-s	0,00%	AX-m	2,27%	TS	7,66%
4	AX-m	0,00%	AX-s	2,33%	AX-m	9,14%
5	AX-s	0,00%	TS	2,36%	AX-s	12,40%
6	TS	0,00%	AO-m	4,99%	AO-m	17,18%
7	SO	0,00%	AO-s	5,01%	AO-s	17,60%

For the problem size 10x2x20, according to the values of the RE_{min} in Table 4, all algorithms were able to find the best-known solutions. Here, $RE_{min} = 0,00\%$, $RE_{avg} \in [2,05\%, 5,01\%]$, $RE_{max} \in [6,67\%, 17,60\%]$. The algorithms implementing ring or random topologies realized on homogeneous islands, as well as random topology realized on all islands have considerably lower RE_{avg} and RE_{max} in comparison with the algorithms exploiting the ring topology built on all islands. Thus, SO, SX and AX-m algorithms outperform the other algorithms while solving the problem of the size 10x2x20.

Table 5 The algorithms ordered non-decreasingly according to their RE_{min}, RE_{avg} and RE_{max} for the size 10x3x20 of the problem Θ_z

Nr	Algm	RE_{min}	Algm	RE_{avg}	Algm	RE_{max}
1	AX-m	0,00%	AX-m	3,61%	AX-m	14,71%
2	SX	0,00%	AX-s	3,73%	SO	15,18%
3	AX-s	0,00%	SO	3,87%	AX-s	15,89%
4	TS	0,00%	SX	4,06%	SX	16,66%
5	SO	0,00%	TS	4,95%	TS	18,28%
6	AO-m	0,00%	AO-s	8,13%	AO-m	24,24%
7	AO-s	0,08%	AO-m	8,13%	AO-s	26,03%

For the problem size 10x3x20, according to the values of the RE_{min} in Table 5, all algorithms, except for AO-s, were able to find the best-known solutions. Here, $RE_{min} \in [0,00\%, 0,08\%]$, $RE_{avg} \in [3,61\%, 8,13\%]$, $RE_{max} \in [14,71\%, 26,03\%]$. The overall results for the size 10x3x20 are nearly the same as for 10x2x20, i.e. the algorithms implementing random or ring topologies realized on homogeneous islands, as well as random topology realized on all islands have considerably lower RE_{avg} and RE_{max} in comparison with the algorithms exploiting the ring topology built on all islands. However, for this size of the problem AX-m algorithm has lower RE_{avg} and RE_{max} than SO. Thus, AX-m, SO and AX-s algorithms outperform the other algorithms while solving the problem of the size 10x3x20.

Table 6 The algorithms ordered non-decreasingly according to their RE_{min}, RE_{avg} and RE_{max} for the size 20x2x20 of the problem Θ_z

Nr	Algm	RE_{min}	Algm	RE_{avg}	Algm	RE_{max}
1	SX	0,00%	AX-m	2,05%	AX-m	9,08%
2	AX-m	0,00%	SO	2,43%	SO	9,76%
3	AX-s	0,00%	AX-s	2,53%	AX-s	9,90%
4	SO	0,00%	AO-m	2,60%	TS	10,81%
5	AO-m	0,00%	AO-s	2,72%	SX	11,00%
6	AO-s	0,00%	SX	2,84%	AO-m	15,19%
7	TS	0,53%	TS	5,09%	AO-s	15,61%

For the problem size 20x2x20, according to the values of the RE_{min} in Table 6, all island-based algorithms were able to find the best-known solutions. Here, $RE_{min} \in [0{,}00\%, 0{,}53\%]$, $RE_{avg} \in [2{,}05\%, 5{,}09\%]$, $RE_{max} \in [9{,}08\%, 15{,}61\%]$. The overall results for the size 20x2x20 are much alike as for 10x2x20, i.e. the algorithms implementing random or ring topologies realized on homogeneous islands, as well as random topology realized on all islands have considerably lower RE_{avg} and RE_{max} in comparison with the algorithms exploiting the ring topology built on all islands. Again, AX-m algorithm has lower RE_{avg} and RE_{max} than SO for this size of the problem. Thus, AX-m, SO and AX-s algorithms outperform the other algorithms while solving the problem of the size 20x2x20.

Table 7 The algorithms ordered non-decreasingly according to their RE_{min}, RE_{avg} and RE_{max} for the size 10x2x50 of the problem Θ_z

Nr	Algm	RE_{min}	Algm	RE_{avg}	Algm	RE_{max}
1	TS	0,00%	AX-m	2,47%	SO	8,69%
2	SX	0,00%	AX-s	2,53%	TS	8,79%
3	AX-m	0,00%	SO	2,77%	SX	9,23%
4	AX-s	0,00%	SX	2,79%	AX-m	11,02%
5	SO	0,00%	TS	3,09%	AX-s	11,19%
6	AO-m	0,00%	AO-m	5,77%	AO-s	15,41%
7	AO-s	0,03%	AO-s	5,78%	AO-m	16,61%

For the problem size 10x2x50, according to the values of the RE_{min} in Table 7, all island-based algorithms, except for AO-s, were able to find the best-known solutions. Here, $RE_{min} \in [0{,}00\%, 0{,}03\%]$, $RE_{avg} \in [2{,}47\%, 5{,}78\%]$, and finally $RE_{max} \in [8{,}69\%, 16{,}61\%]$. The overall results for the size 10x2x50 show that the algorithms implementing random or ring topologies realized on homogeneous islands, as well as random topology realized on all islands have considerably lower RE_{avg} and RE_{max} in comparison with the algorithms exploiting the ring topology built on all islands. For the size 10x2x50 the result do not allow to unequivocally determine the most efficient algorithm, thus we distinguish AX-m, SO, SX and AX-s algorithms as more efficient than AO-m and AO-s algorithms.

For the problem size 10x3x50, according to the values of the RE_{min} in Table 8, all algorithms were able to find the best-known solutions. Here, $RE_{min} \in [0{,}00\%, 0{,}06\%]$, $RE_{avg} \in [3{,}31\%, 5{,}91\%]$, $RE_{max} \in [14{,}66\%, 36{,}35\%]$. The overall results for the size 10x3x50 show that the algorithms implementing random or ring topologies realized on homogeneous islands, as well as random topology realized on all islands have considerably lower RE_{avg} and RE_{max} in comparison with the algorithms exploiting the ring topology built on all islands. For this size of the problem, AX-m, AX-s and SX algorithms outperform the other algorithms while solving the problem.

Table 8 The algorithms ordered non-decreasingly according to their RE_{min}, RE_{avg} and RE_{max} for the size 10x3x50 of the problem Θ_z

Nr	Algm	RE_{min}	Algm	RE_{avg}	Algm	RE_{max}
1	AX-s	0,00%	AX-m	3,31%	AX-m	14,66%
2	SX	0,00%	AX-s	3,46%	AX-s	16,59%
3	AO-s	0,00%	SO	3,86%	SX	17,34%
4	AX-m	0,00%	SX	4,18%	TS	18,16%
5	SO	0,00%	TS	5,72%	AO-s	25,24%
6	AO-m	0,00%	AO-s	5,86%	AO-m	27,02%
7	TS	0,06%	AO-m	5,91%	SO	36,35%

Table 9 The algorithms ordered non-decreasingly according to their RE_{min}, RE_{avg} and RE_{max} for the size 20x2x50 of the problem Θ_z

Nr	Algm	RE_{min}	Algm	RE_{avg}	Algm	RE_{max}
1	SO	0,00%	AX-m	3,84%	SX	11,31%
2	SX	0,00%	AX-s	3,95%	SO	11,77%
3	AX-m	0,00%	SO	5,18%	AX-m	12,44%
4	AX-s	0,00%	SX	5,33%	AX-s	12,49%
5	AO-m	0,87%	TS	6,19%	TS	12,65%
6	AO-s	0,96%	AO-s	7,73%	AO-s	17,45%
7	TS	1,05%	AO-m	7,80%	AO-m	18,76%

For the problem size 20x2x50, according to the values of the RE_{min} in Table 9, all algorithms, except for AO-m and AO-s, were able to find the best-known solutions. Here, we give the intervals of the REs' values: $RE_{min} \in [0{,}00\%, 0{,}96\%]$, $RE_{avg} \in [3{,}84\%, 7{,}80\%]$, $RE_{max} \in [11{,}31\%, 18{,}76\%]$. The overall results for the size 20x2x50 show that the algorithms implementing random or ring topologies realized on homogeneous islands, as well as random topology realized on all islands have considerably lower RE_{avg} and RE_{max} in comparison with the algorithms exploiting the ring topology built on all islands. For the size 20x2x50 the result do not allow to unequivocally determine the most efficient algorithm, thus we distinguish AX-m, SO, SX and AX-s algorithms as more efficient than AO-m and AO-s algorithms.

In order to determine the percentages of the best found solutions for a particular problem size that are of the same quality as the best-known solutions, we had determined the best solutions found within 43 runs of the algorithm for each of 100 problem instances. Next, we counted how many solutions out of obtained 100 had the same quality as the best known for the same problem size and gave this number in percents. The percentages of the best solutions found by the proposed algorithms of the same quality as the best-known solutions are given in Tables 10 - 12.The results in the Tables 10 – 12 confirm unequivocally the previous conclusion, that the algorithms implementing random or ring topologies realized on homogeneous islands, as well as random topology realized on all islands are more efficient that the algorithms exploiting the ring topology built on all islands. Similarly as for REs, AX-m, AX-s and SX prevail other algorithms with clear dominance of AX-m and AX-s, i.e. the algorithms that implement the random topology realized on all islands.

Table 10 The percentage of the best solutions (PBFS), ordered non-increasingly, found by the proposed algorithms that have the same quality as the best-known solutions for the discretisation level $D = 10$

10x2x10	PBFS	10x3x10	PBFS	20x2x10	PBFS
AX-m	69%	AX-s	52%	AX-s	47%
SX	68%	AX-m	49%	AX-m	28%
AX-s	66%	SX	46%	SX	11%
SO	50%	SO	33%	SO	10%
TS	44%	TS	22%	AO-s	2%
AO-s	16%	AO-m	8%	TS	2%
AO-m	8%	AO-s	3%	AO-m	0%

Table 11 The percentage of the best solutions (PBFS), ordered non-increasingly, found by the proposed algorithms that have the same quality as the best-known solutions for the discretisation level $D = 20$

10x2x20	PBFS	10x3x20	PBFS	20x2x20	PBFS
AX-m	36%	AX-m	53%	SX	32%
AX-s	36%	AX-s	37%	AX-m	32%
SX	35%	SX	27%	AX-s	25%
SO	31%	SO	17%	SO	12%
TS	16%	AO-m	9%	AO-m	4%
AO-m	10%	AO-s	9%	AO-s	2%
AO-s	8%	TS	6%	TS	0%

Table 12 The percentage of the best solutions (PBFS), ordered non-increasingly, found by the proposed algorithms that have the same quality as the best-known solutions for the discretisation level $D = 50$

10x2x50	PBFS	10x3x50	PBFS	20x2x50	PBFS
AX-m	42%	AX-m	37%	AX-m	47%
AX-s	30%	AX-s	27%	AX-s	40%
SO	18%	SX	17%	SX	9%
SX	18%	AO-s	12%	SO	3%
TS	4%	SO	10%	TS	1%
AO-m	3%	AO-m	6%	AO-m	0%
AO-s	1%	TS	0%	AO-s	0%

Although, it was possible to determine several most efficient algorithms for each conducted test, we still can't distinguish the most efficient one. In order to do so, we need some universal measure, that could be applied for evaluation of the proposed algorithms. For this reason, we need to transform, or more precisely - normalize RE_{min}, RE_{avg}, RE_{max} and PBFS in a such way, that it would be possible to obtain some estimates that could be aggregated into one estimate, this way enabling the choice of the most efficient algorithm. Because RE and PBFS have opposite evaluation meaning, i.e. the lower RE – the better performance of the algorithm, the lower PBFS – the worse performance of the algorithm, we introduce a new parameter NB = 1 – PBFS instead of PBFS. Thus, let

$$ne_p = \frac{x - x_{min}}{x_{max}} \tag{7}$$

be the formula which we apply to RE_{min}, RE_{avg}, RE_{max} and NB within a paricular problem size in order to obtain a normalized estimate *ne*. In the Equation (7), p – one of the considered parameters, i.e. RE_{min}, RE_{avg}, RE_{max} or NB, x – the value of the considered parameter of the particular algorithm, x_{min}, x_{max} – the minimum or respectively maximum value of the considered parameter within the same problem size among all considered algorithms. After calculating *ne* values for all parameters of all algorithms for all problem sizes, the values obtained for each algorithm were summed into an aggregated estimate. The values of the aggregated estimates were used to make a ranking of the considered algorithms, which is shown in Table 13. As it could be seen in Table 13, the ranking implies the superiority of the algorithms implementing random topology realized on both heterogeneous and homogeneous islands over the algorithms implementing the ring topology. Thus, according to the ranking AX-m algorithm is the most efficient among all considered algorithms.

Table 13 A ranking of the considered algorithms according to the aggregated estimate values

Alg-m	Aggregated estimate	Ranking
AX-m	0,83	1
AX-s	2,05	2
SX	3,06	3
SO	4,94	4
TS	9,61	5
AO-m	13,68	6
AO-s	16,50	7

Table 14 The ranges and deltas of RE_{avg} and RE_{max} values for the considered problem sizes ordered by ΔRE_{avg} and ΔRE_{max} non-decreasingly

prbl. size	RE_{avg} range	ΔRE_{avg}	prbl. size	RE_{max} range	ΔRE_{max}
20x2x20	2,05% - 5,09%	2,59%	20x2x20	9,08% - 15,61%	6,53%
10x3x50	3,31% - 5,91%	2,60%	20x2x10	11,47% - 18,19%	6,72%
10x2x20	2,05% - 5,01%	2,96%	10x2x10	9,76% - 16,53%	6,77%
10x2x10	3,28% - 6,30%	3,02%	20x2x50	11,31% - 18,76%	7,45%
10x2x50	2,47% - 5,78%	3,31%	10x2x50	8,69% - 16,61%	7,92%
20x2x50	3,84% - 7,80%	3,96%	10x3x10	15,92% - 26,06%	10,14%
10x3x10	4,67% - 8,71%	4,04%	10x2x20	6,67% - 17,60%	10,93%
10x3x20	3,61% - 8,13%	4,52%	10x3x20	14,71% - 26,03%	11,32%
20x2x10	4,76% - 9,31%	4,64%	10x3x50	14,66% - 36,35%	21,69%

Table 15 The range and delta of PBFS values for the considered problem sizes ordered by ΔPBFS non-decreasingly

prbl. size	PBFS range	ΔPBFS
10x2x20	8% - 36%	28%
20x2x20	2% - 32%	30%
10x3x50	6% - 37%	31%
10x2x50	1% - 42%	41%
10x3x20	9% - 53%	44%
20x2x10	0% - 47%	47%
20x2x50	0% - 47%	47%
10x3x10	3% - 52%	49%
10x2x10	8% - 69%	61%

As it could be seen from the experimental results described above, it is possible to reduce the REs of the solutions found by the considered algorithms just by changing the interconnection topology of the constituent islands. The Table 14 shows that by changing the interconnection topology RE_{avg} can be reduced by 2,59% - 4,64% and RE_{max} by 6,53% - 21,69% dependently on the problem size. The RE_{avg} and RE_{max} ranges were taken from the Tables 1 – 9. Similarly, the Table 15 shows that the percentage of the best found solutions that have the same quality as the best-known solutions can be increased by 28% - 61% dependently on the problem size. The PBFS ranges were taken from the Tables 10 – 12.

Finally, in Tables 16 – 17, we observe the influence of the level of the continuous resource discretisation D on the REs of the found solutions. In Table 16, for the problem size 10x2xD, $D \in \{10, 20, 50\}$, for almost all algorithms except for AO-s, both RE_{min} and RE_{avg} have the lowest values when $D = 20$. Thus, the influence of D on the REs for the considered problem size could generalized by the following relations: $RE_{min/avg}(D = 20) < RE_{min/avg}(D = 50) < RE_{min/avg}(D = 10)$. For the RE_{max}, the results are mixed and it's impossible to derive one clear rule for all algorithms. The influence of the discretisation level D on the REs of the solutions found by the considered algorithms for the problem size 10x3xD, $D \in \{10, 20, 50\}$, according to the Table 17 could be described generally by the following relations: $RE_{min}(D = 20) \leq RE_{min}(D = 50) < RE_{min}(D = 10)$, except for AO-s, and $RE_{avg}(D = 50) < RE_{avg}(D = 20) < RE_{avg}(D = 10)$, except for SX and TS. For the RE_{max}, the results are mixed and it's impossible to derive one clear rule for all algorithms. The influence of the discretisation level D on the REs of the solutions found by the considered algorithms for the problem size 20x2xD, $D \in \{10, 20, 50\}$, according to the Table 18 could be described generally by the following relations: $RE_{min}(D = 20) \leq RE_{min}(D = 50) < RE_{min}(D = 10)$ except for TS, and $RE_{avg}(D = 20) < RE_{avg}(D = 50) < RE_{avg}(D = 10)$. For the RE_{max}, the results are mixed and it's impossible to derive one clear rule for all algorithms. Below we tabularise the obtained relations together in Table 19:

Table 16 The influence of the level of the continuous resource discretisation D on the REs of the found solutions for the problem size 10x2xD, $D \in \{10, 20, 50\}$

Algm	RE_{min}			RE_{avg}			RE_{max}		
	10	20	50	10	20	50	10	20	50
AO-m	0,01%	0,00%	0,00%	6,30%	4,99%	5,77%	14,68%	17,18%	16,61%
AO-s	0,01%	0,00%	0,03%	6,30%	5,01%	5,78%	16,53%	17,60%	15,41%
AX-m	0,01%	0,00%	0,00%	3,54%	2,27%	2,47%	12,33%	9,14%	11,02%
AX-s	0,01%	0,00%	0,00%	3,58%	2,33%	2,53%	11,39%	12,40%	11,19%
SO	0,19%	0,00%	0,00%	3,28%	2,05%	2,77%	9,76%	6,67%	8,69%
SX	0,01%	0,00%	0,00%	3,31%	2,07%	2,79%	10,00%	7,04%	9,23%
TS	0,01%	0,00%	0,00%	3,49%	2,36%	3,09%	9,91%	7,66%	8,79%

Table 17 The influence of the level of the continuous resource discretisation D on the REs of the found solutions for the problem size 10x3xD, $D \in \{10, 20, 50\}$

Algm	RE_{min}			RE_{avg}			RE_{max}		
	10	20	50	10	20	50	10	20	50
AO-m	0,10%	0,00%	0,00%	8,66%	8,13%	5,91%	26,06%	24,24%	27,02%
AO-s	0,28%	0,08%	0,00%	8,71%	8,13%	5,86%	25,67%	26,03%	25,24%
AX-m	0,00%	0,00%	0,00%	4,68%	3,61%	3,31%	16,07%	14,71%	14,66%
AX-s	0,00%	0,00%	0,00%	4,74%	3,73%	3,46%	19,33%	15,89%	16,59%
SO	0,00%	0,00%	0,00%	4,67%	3,87%	3,86%	15,92%	15,18%	36,35%
SX	0,01%	0,00%	0,00%	4,69%	4,06%	4,18%	18,39%	16,66%	17,34%
TS	0,07%	0,00%	0,06%	5,36%	4,95%	5,72%	17,72%	18,28%	18,16%

Table 18 The influence of the level of the continuous resource discretisation D on the REs of the found solutions for the problem size 20x2xD, $D \in \{10, 20, 50\}$

Algm	RE_{min}			RE_{avg}				RE_{max}		
	10	20	50	10	20	50		10	20	50
AO-m	1,12%	0,00%	0,87%	9,31%	2,60%	7,80%	215	16,71%	15,19%	18,76%
AO-s	1,16%	0,00%	0,96%	9,19%	2,72%	7,73%	251	18,19%	15,61%	17,45%
AX-m	0,00%	0,00%	0,00%	4,76%	2,05%	3,84%	215	11,81%	9,08%	12,44%
AX-s	0,26%	0,00%	0,00%	4,86%	2,53%	3,95%	251	12,54%	9,90%	12,49%
SO	0,51%	0,00%	0,00%	5,46%	2,43%	5,18%	215	11,47%	9,76%	11,77%
SX	0,40%	0,00%	0,00%	5,56%	2,84%	5,33%	251	11,74%	11,00%	11,31%
TS	0,68%	0,53%	1,05%	6,26%	5,09%	6,19%	215	12,09%	10,81%	12,65%

Table 19 The relations among the REs on different discretisation levels D, $D \in \{10, 20, 50\}$, for the considered problem sizes

Prbl.size	Relations among the REs on different discretisation levels D, $D \in \{10, 20, 50\}$
10x2xD	$RE_{min}(D = 20) < RE_{min}(D = 50) < RE_{min}(D = 10)$
	$RE_{avg}(D = 20) < RE_{avg}(D = 50) < RE_{avg}(D = 10)$
	RE_{max} – mixed
10x3xD	$RE_{min}(D = 20) \bullet RE_{min}(D = 50) < RE_{min}(D = 10)$, except for AO-s
	$RE_{avg}(D = 50) < RE_{avg}(D = 20) < RE_{avg}(D = 10)$, except for SX and TS
	RE_{max} – mixed
20x2xD	$RE_{min}(D = 20) \bullet RE_{min}(D = 50) < RE_{min}(D = 10)$ except for TS
	$RE_{avg}(D = 20) < RE_{avg}(D = 50) < RE_{avg}(D = 10)$
	RE_{max} – mixed

As it could be seen in Table 19, it's impossible to determine unequivocally the discretisation level on which REs of the found solutions are the lowest. However, it could be pointed at the relation REs(D = 20) < REs(D = 50) < REs(D = 10) as the most frequent relation. This might impose the conclusion, that the high discretisation level does not ensure the lowest values of the REs and the additional research is needed to identify the most appropriate discretisation of the continuous resource.

5 Conclusion

In the chapter, we consider the population learning algorithm (PLA2), earlier designed by the authors for solving the problem of scheduling non-preemtable tasks on parallel identical machines under constraint on discrete resource and requiring, additionally, renewable continuous resource to minimize the schedule length. The PLA2 model can be also viewed as an island model, were homogeneous as well as heterogeneous islands are connected to each other according to some topology and exchange individuals in order to collectively find best possible solution to the problem. The PLA2's island-based design can be easily used to construct an agent system with cooperating agents. The main goal of our research was to find out whether a topology of a learning stages (or islands), might have some effect on the algorithm's efficiency. For this reason we proposed six versions of PLA2 that differs from each other by their structure and migration scheme. The most important conclusion that can be drawn from the experimental results is that the interconnection topology of the constituent islands might have a noticeable impact on the quality of the solutions yielded by PLA2. It is possible to reduce the relative errors of the solutions found by PLA2 by order of 2,59% - 4,64% for RE_{avg} and 6,53% - 21,69% for RE_{max} dependently on the problem size. Similarly, the percentage of the best found solutions that have the same quality as the best-known solutions can be increased dependently on the problem size by 28% - 61%. The ranking of the considered algorithms that was designed to reveal the most efficient interconnection topology implies the superiority of the algorithms implementing random topology realized on all available islands, i.e. heterogeneous and homogeneous, or exclusively on homogeneous islands, over the algorithms implementing the ring topology. However, the algorithm implementing the directed ring topology realized exclusively on homogeneous islands for some problem sizes yielded solutions that had the lowest RE_{avg} and RE_{max} values. It should be mentioned here, that all the conclusions are valid for particular implementations of the algorithms used in the experiments. The values of some parameters of the algorithms were determined during their tuning and should be determined on the way of an exhaustive experiment. Our further research should concern other parameters and interconnection topologies that might allow to improve the efficiency of the algorithms that implement the island-based model.

References

1. De Boer, P.-T., Kroese, D.P., Mannor, S., Rubinstein, R.Y.: A Tutorial on the Cross-Entropy Method. Annals of Operations Research 134(1), 19–67 (2005)
2. Czarnowski, I., Gutjahr, W.J., Jędrzejowicz, P., Ratajczak, E., Skakovski, A., Wierzbowska, I.: Scheduling Multiprocessor Tasks in Presence of Correlated Failures. Central European Journal of Operations Research 11(2), 163–182 (2003); Luptaćik, M., Wildburger, U.L. (eds.) Physika-Verlag, A Springer-Verlag Company, Heidelberg
3. Jędrzejowicz, J., Jędrzejowicz, P.: Population–Based Approach to Multiprocessor Task Scheduling in Multistage Hybrid Flowshops. In: Palade, V., Howlett, R.J., Jain, L. (eds.) KES 2003. LNCS (LNAI), vol. 2773, pp. 279–286. Springer, Heidelberg (2003)
4. Jędrzejowicz, J., Jędrzejowicz, P.: PLA–Based Permutation Scheduling. Foundations of Computing and Decision Sciences 28(3), 159–177 (2003)
5. Jędrzejowicz, J., Jędrzejowicz, P.: New Upper Bounds for the Permutation Flowshop Scheduling Problem. In: Ali, M., Esposito, F. (eds.) IEA/AIE 2005. LNCS (LNAI), vol. 3533, pp. 232–235. Springer, Heidelberg (2005)
6. Jędrzejowicz, P.: Social Learning Algorithm as a Tool for Solving Some Difficult Scheduling Problems. Foundation of Computing and Decision Sciences 24, 51–66 (1999)
7. Jędrzejowicz, P., Skakovski, A.: A Population Learning Algorithm for Discrete-Continuous Scheduling with Continuous Resource Discretisation. In: Chen, Y., Abraham, A. (eds.) 6th International Conference on Intelligent Systems Design and Applications, ISDA 2006 Special session: Nature Imitation Methods Theory and Practice (NIM 2006), October 16-18, vol. II, pp. 1153–1158. IEEE Computer Society, Jinan (2006)
8. Jędrzejowicz, P., Skakovski, A.: A Cross-Entropy Based Population Learning Algorithm for Discrete-Continuous Scheduling with Continuous Resource Discretisation. In: Lovrek, I., Howlett, R.J., Jain, L.C. (eds.) KES 2008, Part I. LNCS (LNAI), vol. 5177, pp. 82–89. Springer, Heidelberg (2008)
9. Józefowska, J., Węglarz, J.: On a methodology for discrete-continuous scheduling. European J. Oper. Res. 107(2), 338–353 (1998)
10. Józefowska, J., Mika, M., Różycki, R., Waligóra, G., Węglarz, J.: Solving discrete-continuous scheduling problems by Tabu Search. In: 4th Metaheuristics International Conference MIC 2001, Porto, Portugal, July 16-20, pp. 667–671 (2001)
11. Józefowska, J., Różycki, R., Waligóra, G., Węglarz, J.: Local search metaheuristics for some discrete-continuous scheduling problems. European J. Oper. Res. 107(2), 354–370 (1998)
12. Różycki, R.: Zastosowanie algorytmu genetycznego do rozwiązywania dyskretno-ciągłych problemów szeregowania. PhD dissertation, Istitute of Computing Science, Poznań University of Technology, Piotrowo 3A, 60-965, Poznań, Poland (2000)
13. Rubinstein, R.Y.: Optimization of computer simulation models with rare events. European Journal of Operations Research 99, 89–112 (1997)

Triple-Action Agents Solving the MRCPSP/Max Problem

Piotr Jędrzejowicz* and Ewa Ratajczak-Ropel

Abstract. In this chapter the A-Team architecture for solving the multi-mode resource-constrained project scheduling problem with minimal and maximal time lags (MRCPSP/max) is proposed and experimentally validated. To solve this problem an asynchronous team of agents implemented using JABAT middleware has been proposed. Four kinds of optimization agent has been used. Each of them acts in three ways depending whether the received initial solution is feasible or not. Computational experiment involves evaluation of optimization agents performance within the A-Team. The chapter contains the MRCPSP/max problem formulation, description of the proposed architecture for solving the problem instances, description of optimization algorithms, description of the experiment and the discussion of the computational experiment results.

1 Introduction

The chapter proposes an agent-based approach to solving instances of the MRCPSP/max, known also in the literature as the MRCPSP-GPR problem. MRCPSP stands for the Multi-mode Resource-Constrained Project Scheduling Problem, max or GPR is used to describe precedence relations as minimal and maximal time lags, also called Generalized Precedence Relations (GPR) or temporal constraints or time windows. MRCPSP/max has attracted a lot of attention and many exact and heuristic algorithms have been proposed for solving it (see for example [20], [9], [10], [18], [3]).

MRCPSP/max is a generalization of the RCPSP/max and thus it is NP-hard [2]. The approaches to solve this problem produce either approximate solutions or can

Piotr Jędrzejowicz · Ewa Ratajczak-Ropel
Department of Information Systems, Gdynia Maritime University,
Morska 83, 81-225 Gdynia, Poland
e-mail: {pj,ewra}@am.gdynia.pl

* Corresponding author.

I. Czarnowski et al. (Eds.): Agent-Based Optimization, SCI 456, pp. 103–122.
DOI: 10.1007/978-3-642-34097-0_5

be applied for solving instances of the limited size. Hence, searching for more effective algorithms and solutions to the MRCPSP/max problem is still a lively field of research. One of the promising directions of such research is to take advantage of the parallel and distributed computation solutions, which are the feature of the contemporary multiple-agent systems.

The multiple-agent systems are an important and intensively expanding area of research and development. There is a number of multiple-agent approaches proposed to solve different types of optimization problems. One of them is the concept of an asynchronous team (A-Team), originally introduced in [21]. The A-Team paradigm was used to develop the JADE-based environment for solving a variety of computationally hard optimization problems called E-JABAT [1]. E-JABAT is a middleware supporting the construction of the dedicated A-Team architectures based on the population-based approach. The mobile agents used in E-JABAT allow for decentralization of computations and use of multiple hardware platforms in parallel, resulting eventually in more effective use of the available resources and reduction of the computation time.

In this chapter the E-JABAT-based A-Team architecture for solving the MRCPSP/max problem instances is proposed and experimentally validated. A-Team includes optimization agents which represent heuristic algorithms. The proposed approach is an extension and improvement of the A-Team described in [13]. A new kind of optimization agent was added and all agents were redefined in order to act in three ways depending on feasibility or unfeasibility of the initial solution.

Section 2 of the chapter contains the MRCPSP/max problem formulation. Section 4 provides details of the E-JABAT architecture implemented for solving the MRCPSP/max problem instances. In section 5 the computational experiment is described. In section 6 the computational experiment results are presented. Section 7 contains conclusions and suggestions for future research.

2 Problem Formulation

In the multi-mode resource-constrained project scheduling problem with minimal and maximal time lags (MRCPSP/max) a set of $n+2$ activities $V = \{0, 1, \ldots, n, n+1\}$ is considered. Each activity has to be processed without interruption to complete the project. The dummy activities 0 and $n+1$ represent the beginning and the end of the project. For each activity $i \in V$ a set $M_i = \{1, ..., |M_i|\}$ of (execution) modes is available. Activities 0 and $n+1$ can be performed in only one mode: $M_0 = M_{n+1} = \{1\}$. Each activity $i \in V$ has to be performed in exactly one mode $m_i \in M_i$. The duration (processing time) of an activity i, $i \in V$ executed in m_i mode is denoted by d_{im_i}, $d_{im_i} \in Z_{\geq 0}$. The processing times of activity 0 and $n+1$ equals 0, i.e. $d_{00} = d_{n+1\,0} = 0$.

S_i and C_i stand for the start time and the completion time (of the performance) of activity i, respectively. If we define $S_0 = 0$, S_{n+1} stands for the project duration. Provided that activity i starts in mode m_i at time S_i, it is being executed at each point in time $t \in [S_i, S_i + d_{im_i})$.

Between the start time S_i of activity i, which is performed in mode $m_i \in M_i$, and the start time S_j of activity j $(i \neq j)$, which is performed in mode $m_j \in M_j$, a minimum time lag $d^{min}_{im_i,jm_j} \in Z_{\geq 0}$ or a maximum time lag $d^{max}_{im_i,jm_j} \in Z_{\geq 0}$ can be given. Note, that a time lag between activity i and activity j depends on mode m_i as well as on mode m_j.

Activities and time lags are represented by an activity-on-node (AoN) network $N = \langle V, E, \delta \rangle$ with node set V, arc set E, and arc weight function δ. Each element of node set V represents an activity. In the following, we do not distinguish between an activity and the corresponding node. An arc $\langle i, j \rangle \in E$ indicates that a time lag between S_i and S_j has to be observed. Arc weight function δ assigns to each arc $\langle i, j \rangle \in E$ a $|M_i| \times |M_j|$ matrix of arc weights as follow: for a minimum time lag $d^{min}_{im_i,jm_j}$ we set $\delta_{im_i,jm_j} = d^{min}_{im_i,jm_j}$, and for a maximum time lag $d^{max}_{im_i,jm_j}$ we set $\delta_{im_i,jm_j} = -d^{max}_{im_i,jm_j}$.

There are the set of renewable resources R^R and the set of nonrenewable resources R^N considered in this problem, $|R^R|, |R^N| \in Z_{>0}$. The availability of each renewable resource type $k \in R^R$ in each time period is R^R_k units. The availability of each nonrenewable resource type $k \in R^N$ is R^N_k units in total. Provided that activity i is performed in mode m_i, $r^R_{im_ik}$ units of renewable resource $k \in R^R$ are used at each point in time at which activity i is being executed. Moreover, $r^N_{im_ik}$ units of nonrenewable resource $k \in R^N$ are consumed in total. For activities 0 and $n+1$ we set $r_{01k} = r_{n+11k} = 0$ for $k \in R^R$ and $r^N_{01k} = r^N_{n+10k} = 0$ for $k \in R^N$.

The solution of the problem is a schedule (M, S) consisting of the mode vector M and a vector of activities starting times $S = [S_0, \ldots, S_{n+1}]$, where $S_0 = 0$ (project always begins at time zero). The mode vector assigns to each activity $i \in V$ exactly one mode $m_i \in M_i$ - execution mode of activity i. The start time vector S assigns to each activity $i \in V$ exactly one point in time as start time S_i where $S_0 = 0$ (project always begins at time zero). Precedence relations are described by the following formula: $S.S_j - S.S_i \geq \delta_{im_i,jm_j}$, where $\langle i, j \rangle \in E$.

The objective is to find a schedule (M, S) where precedence and resource constraints are satisfied, such that the schedule duration $T(S) = S_{n+1}$ is minimized. The detailed description of the problem can be found in [10] or [3]. The MRCPSP/max, as an extension of the RCPSP and RCPSP/max, belongs to the class of NP-hard optimization problems [2].

3 E-JABAT Environment

E-JABAT is a middleware allowing to design and implement A-Team architectures for solving various combinatorial optimization problems, such as the resource-constrained project scheduling problem (RCPSP), the traveling salesman problem (TSP), the clustering problem (CP), the vehicle routing problem (VRP). It has been implemented using JADE framework. Detailed information about E-JABAT and its implementations can be found in [11] and [1]. The problem-solving paradigm on which the proposed system is based can be best defined as the population-based approach.

E-JABAT produces solutions to combinatorial optimization problems using a set of optimization agents, each representing an improvement algorithm. Each improvement (optimization) algorithm when supplied with a potential solution to the problem at hand, tries to improve this solution. An initial population of solutions (individuals) is generated or constructed. Individuals forming an initial population are, at the following computation stages, improved by independently acting agents. Main functionality of the proposed environment includes organizing and conducting the process of search for the best solution.

To perform the above described cycle two main classes of agents are used. The first class called OptiAgent is a basic class for all optimization agents. The second class called SolutionManager is used to create agents or classes of agents responsible for maintenance and updating individuals in the common memory. All agents act in parallel. Each OptiAgent represents a single improvement algorithm (for example: local search, simulated annealing, tabu search, genetic algorithm etc.).

Other important classes in E-JABAT include: Task representing an instance or a set of instances of the problem and Solution representing the solution. To initialize the agents and maintain the system the TaskManager and PlatformManager classes are used. Objects of the above classes also act as agents. Up to now the E-JABAT environment has been used to solve instances of the following problems: the resource-constrained project scheduling problem (RCPSP), the traveling salesman problem (TSP), the clustering problem (CP), the vehicle routing problem (VRP).

E-JABAT environment has been designed and implemented using JADE (Java Agent Development Framework), which is a software framework supporting the implementation of multi-agent systems. More detailed information about E-JABAT environment and its implementations can be found in [11] and [1].

4 E-JABAT for Solving the MRCPSP/max Problem

E-JABAT environment was successfully used for solving the RCPSP, MRCPSP and RCPSP/max problems [12]. In the proposed approach the agents, classes describing the problem and ontologies have been implemented for solving the discussed problem. The above forms the package called JABAT.MRCPSPmax.

Classes describing the problem are responsible for reading and preprocessing the data and generating random instances of the problem. The discussed set includes the following classes:

- MRCPSPmaxTask inheriting from the Task class and representing the instance of the problem,
- MRCPSPmaxSolution inheriting from the Solution class and representing the solution of the problem instance,
- Activity representing the activity of the problem,
- Mode representing the activity mode,
- Resource representing the renewable or nonrenewable resource,
- PredSuccA representing the predecessor or successor of the activity.

- PredSuccM and PredSuccT representing the matrix of arc weights. The matrix is needed to describe time lags for all pairs of modes of each two activities connected by the arc in the network N.

The next set includes classes allowing for definition of the vocabulary and semantics for the content of messages exchange between agents. In the proposed approach the messages include all data representing the task and solution. The discussed set includes the following classes:

- MRCPSPmaxTaskOntology inheriting from the TaskOntology class,
- MRCPSPmaxSolutionOntology inheriting from the SolutionOntology class,

The last set includes classes describing the optimization agents. Each of them includes the implementation of an optimization heuristic used to solve the problem. All of them are inheriting from OptiAgent class. The set includes:

- optiLSAm denoting Local Search Algorithm (TA_LSAm),
- optiLSAe denoting Local Search Algorithm (TA_LSAe),
- optiTSAe denoting Tabu Search Algorithm (TA_TSAe),
- optiCA denoting Crossover Algorithm (TA_CA),
- optiPRA denoting Path Relinking Algorithm (TA_PRA),

The proposed approach and algorithms are based on the LSA, CA and PRA described in [13] and double-action agents DA_LSA, DA_TSA, DA_CA and DA_PRA described in [14]. However, the above algorithms have been modified, extended and, in addition, equipped with the ability to undertake the third action as describe later. The local search algorithm has been implemented in two versions TA_LSAm and TA_LSAe, which differ in the kind of the move used.

The triple-action optimization agents (TA_) use their algorithms to solve MRCPSP/max problem instances. Each optimization agent needs one (in the case of TA_LSAm, TA_LSAe and TA_TSAe) or two (TA_CA and TA_PRA) initial solutions. In many instances of MRCPSP/max problem it is difficult to generate a population of feasible solutions or even finding a single feasible solution. On the other hand the proposed LSAm, LSAe, TSAe, CA and PRA algorithms are seldom effective when used with unfeasible solutions. There are two main reasons of the above difficulty: lack of the nonrenewable resources or/and presence of cycles in the respective AoN network. The proposed triple-action agents deal with the problem through applying the standard optimization procedure if, at least one feasible initial solution has been found. If not, the nonrenewable resources are checked and if there is a lack of them the second action is initiated. If the second action does not produce a feasible solution the third action is initiated. An agent tries to find a schedule with the minimal cycle time where cycle time is calculated as the sum of elements over the diagonal of the longest path matrix. The matrix is calculated using Floyd-Warshall triple algorithm described in [17].

The objective of the proposed approach is to find the best feasible schedule (M,S). The procedure of finding a new solution from the schedule (M,S) is based on the SGSU (Serial Generation Scheme with Unscheduling) described in [17] with several different priority rules.

Pseudo-codes of the algorithms are shown in Figures 1, 2, 3, 4, 5, respectively. In pseudo-codes S denotes the schedule (M,S) with ordered list of activities. The solution is calculated using procedure based on SGSU. The objective functions used are as follow:

```
int objectiveFunctionS(S)
    { return S.S_{n+1} }
int objectiveFunctionN(S)
    { return quantities of nonrenewable resources R^N used by S }
int objectiveFunctionF(S)
    { return cycles time for S }
```

In the pseudo-codes S denotes the schedule (M,S) with ordered list of activities. All presented algorithms can use initial schedules which does not necessarily guarantee obtaining feasible solutions.

The LSAm (Figure 1) is a local search algorithm which finds local optimum by moving chosen activity with each possible mode to all possible places in the schedule. For each combination of activities the value of possible solution is calculated. The best schedule is returned. The parameter *iterationNumber* means a maximum number of iterations in which no improvement is found. Two procedures are used to make the moves. The makeMove(S,pi,pj,m_{pi}^{new}) means moving the activity form position pi in the schedule (activity list) S to position pj and simultaneously changing the current activity mode m_{pi} to m_{pi}^{new}. The reverseMove(S,pi,pj,m_{pi}^{new}) means canceling the move i.e. moving the activity from position pj to position pi and simultaneously changing the activity mode from m_{pi}^{new} to previous one m_{pi}. The activities positions in the schedule are chosen using *step* parameter. The best schedule is remembered and finally returned.

The LSAe (Figure 2) is a local search algorithm which finds local optimum by exchanging chosen activity with each possible mode with other chosen activity in the schedule. For each combination of activities the value of possible solution is calculated. The best schedule is returned. The parameter *iterationNumber* defines a maximum number of iterations in which no improvement is found. Two procedures are used to make the moves. The makeExchange(S,pi,m_{pi}^{new},pj,m_{pj}^{new}) means exchanging the activity form position pi in the schedule (activity list) S with activity from position pj and simultaneously changing the chosen activity modes m_{pi} to m_{pi}^{new} and m_{pj} to m_{pj}^{new}. The reverseExchange(S,pi,nm_i,pj,nm_j) means cancelling the exchange. The activities positions in the schedule are chosen using *step* parameter. The best schedule is remembered and finally returned.

The TSAe (Figure 3) is an implementation of the tabu search metaheuristic [5], [6], [7]. In a schedule the pairs of activities and simultaneously modes of these activities are changed. The parameter *iterationNumber* denotes a maximum number of iterations in which no improvement is found. Two procedures are used to make the moves. The makeExchange(S,pi,m_{pi}^{new},pj,m_{pj}^{new}) and reverseExchange procedures are the same as in the case of LSAe algorithm. The activities in the schedule are chosen using *step* parameter. Selected moves are remembered in a tabu list. For example:

```
TA_LSAm(initialSchedule)
{
    S = initialSchedule
    if(S is not feasible)
        if(there is a lack of nonrenewable resources in S)
            S =LSAm(S, startActPosN, itNumN, stepN, objectiveFunctionN)
        if(there are cycles in S)
            S =LSAm(S, startActPosF, itNumF, stepF, objectiveFunctionF)
    bestS =LSAm(S, startActPos, itNum, step, objectiveFunctionS)
    return bestS
}

LSAm(S, startActivityPosition, iterationNumber, step, objectiveFunction)
{
    it = iterationNumber
    bestS = S
    pi = startActivityPosition
    while(it>0)
    {
        bestSit = S
        pi = ++pi%(n-2)+1
        for(pj = pi+step; pj < n-1; pj = pj+step)
        {
            for(all modes m_pi^new in activity from position pi in S)
            {
                makeMove(S, pi, pj, m_pi^new)
                if(S is better than bestSit due to objectiveFunction)
                    bestSit = S
                reverseMove(S, pi, pj, m_pi^new)
            }
        }
        if(bestSit is better than bestS due to objectiveFunction)
        {
            bestS = bestSit
            it = iterationNumber
        }
        else it-
    }
    return bestS
}
```

Fig. 1 Pseudo-codes of the TA_LSAm and LSAm algorithms

- Making tabu the exchange move (pi,m_{pi},m_{pi}^{new},pj,m_{pj},m_{pj}^{new},*iterationNumber*) prevents from repeating the same move. It block all moves that exchange activity from position pi with activity from position pj and simultaneously change their modes from m_{pi} to m_{pi}^{new} and from m_{pj} to m_{pj}^{new} for all iterations *iterationNumber*;
- Making tabu the exchange move (null,null,null,pi,m_{pi},null,10) block exchanges of any activity with activity from position pi performed in mode m_{pi} for 10 iterations.

```
TA_LSAe(initialSchedule)
{  S = initialSchedule
   if(S is not feasible)
      if(there is a lack of nonrenewable resources in S)
         S=LSAe(S, startActPosN, itNumN, stepN, objectiveFunctionN)
      if(there are cycles in S)
         S=LSAe(S, startActPosF, itNumF, stepF, objectiveFunctionF)
   bestS=LSAe(S, startActPos, itNum, step, objectiveFunctionS)
   return bestS
}

LSAe(S, startActivityPosition, iterationNumber, step, objectiveFunction)
{
   it = iterationNumber
   bestS = S
   pi = startActivityPosition
   while(it>0)
   {
      bestSit = S
      pi = ++pi%(n-2)+1
      for(pj=pi+step; pj<n-1; pj=pj+step)
      {
         for(all modes m_pi^new in activity from position pi in S)
            for(all modes m_pj^new in activity from position pj in S)
            {
               makeExchange(S, pi, m_pi^new, pj, m_pj^new)
               if(S is better than bestSit due to objectiveFunction)
                  bestSit = S
               reverseExchange(S, pi, m_pi^new, pj, m_pj^new)
            }
      }
      if(bestSit is better than bestS due to objectiveFunction)
      {
         bestS = bestSit
         it = iterationNumber
      }
      else it-
   }
   return bestS
}
```

Fig. 2 Pseudo-codes of the TA_LSAe and LSAe algorithms

The best schedule is remembered and finally returned.

The CA (Figure 4) is an algorithm based on the idea of the one point crossover operator. For a pair of solutions one point crossover is applied. The *step* argument determines the frequency the operation is performed. The `makeCrossover`(S, $S1$, $S2$, cp) constructs the S schedule using one point crossover operator to two initial schedules $S1$ and $S2$ with crossover point cp. The best schedule is remembered and finally returned.

```
TA_TSAe(initialSchedule)
{
   S = initialSchedule
   if(S is not feasible)
      if(there is a lack of nonrenewable resources in S)
         S =TSAe(S, startActPosN, itNumN, stepN, objectiveFunctionN)
      if(there are cycles in S)
         S =TSAe(S, startActPosF, itNumF, stepF, objectiveFunctionF)
   bestS =TSAe(S, startActPos, itNum, step, objectiveFunctionS)
   return bestS
}

TSAe(S, startActivityPosition, iterationNumber, step, objectiveFunction)
{
   TabuList = ∅
   it = iterationNumber
   bestS = S
   pi = startActivityPosition
   while(it>0)
   {
      bestSit =null
      for(pj = pi+1; pi < n−1; pj = pj+step)
         for(all modes in activity from position pi in S)
            for(all modes in activity from position pj in S)
            {
               move = (S, pi, m_pi^new, pj, m_pj^new)
               if(move is not in TabuList or
                  is better than bestS due to objectiveFunction)
               {
                  makeExchange(move)
                  if(S is better than bestSit due to objectiveFunction)
                     bestSit = S
                  reverseExchange(move)
               }
            }
      update TabuList
      if(bestSit is not null)
      {
         if(bestSit is better than bestS due to objectiveFunction)
         {
            bestS = bestSit
            it = iterationNumber
         }
         add moves to TabuList:
            (pi, m_pi, m_pi^new, pj, m_pj, m_pj^new, iterationNumber)
            (null,null,null, pi, m_pi, null,10)
            (null,null,null, pi, m_pi^new, null,10)
            (pi, m_pi, null,null,null,null,10)
            (pi, m_pi^new, null,null,null,null,10)
            (pj, null,null, pi, null,null, iterationNumber/2)
      }
      else it−
      pi = pi%(n−2) + step
   }
   return bestS
}
```

Fig. 3 Pseudo-codes of the TA_TSAe and TSAe algorithms

```
TA_CA(initialSchedule1, initialSchedule2, step)
{
   S1 = initialSchedule1;  S2 = initialSchedule2
   bestS = better from S1 and S2 due to objectiveFunctionS
   if(bestS is not feasible
      and there is a lack of nonrenewable resources in it)
   {
      S =CA(S1, S2, stepN, objectiveFunctionN)
      if(S is better than worse from S1 and S2 due to objectiveFunctionS)
         exchange worse from S1 and S2 due to objectiveFunctionS for S
   }
   bestS = better from S1 and S2 due to objectiveFunctionS
   if(bestS is not feasible)
   {
      S =CA(S1, S2, stepF, objectiveFunctionF)
      if(S is better than worse from S1 and S2 due to objectiveFunctionS)
         exchange worse from S1 and S2 due to objectiveFunctionS for S
   }
   bestS =CA(S1, S2, step, objectiveFunctionS)
   return bestS
}

CA(S1, S2, step, objectiveFunction)
{
   bestS = better from S1 and S2 due to objectiveFunction
   bestS = S
   for(cp = 1;  cp < n;  cp+=step)
   {
      makeCrossover(S, S1, S2, cp)
      for(all modes of activities in crossover point cp
         S = best schedule due to objectiveFunction
      if(S is better than bestS due to objectiveFunction) bestS = S
   }
   return bestS
}
```

Fig. 4 Pseudo-codes of the TA_CA and CA algorithms

The PRA (Figure 5) is an implementation of the path-relinking algorithm [4], [8]. For a pair of solutions a path between them is constructed. The path consists of schedules obtained by carrying out a single move from the preceding schedule. The move is understood as moving one of the activities to a new position simultaneously changing its mode. For each schedule in the path the value of the respective solution is checked. The best schedule is remembered and finally returned. In the PRA the same `makeMove` procedure is used as in the case of LSAm algorithm.

All optimization agents co-operate together using the E-JABAT common memory. The initial population in the common memory is generated randomly with the exception of a few individuals which are generated by heuristics based on the priority rules [17] and procedure based on SGSU. Because it is difficult to obtain feasible solution for some MRCPSP/max problem instances, the random drawing of an

```
TA_PRA(initialSchedule1, initialSchedule2)
{
   S1 = initialSchedule1;  S2 = initialSchedule2
   bestS = better from S1 and S2 due to objectiveFunctionS
   if(bestS is not feasible
      and there is a lack of nonrenewable resources in it)
   {
      S =PRA(S1, S2, objectiveFunctionN)
      if(S is better than worse from S1 and S2 due to objectiveFunctionS)
         exchange worse from S1 and S2 due to objectiveFunctionS for S
   }
   bestS = better from S1 and S2 due to objectiveFunctionS
   if(bestS is not feasible)
   {
      S =PRA(S1, S2, objectiveFunctionF)
      if(S is better than worse from S1 and S2 due to objectiveFunctionS)
         exchange worse from S1 and S2 due to objectiveFunctionS for S
   }
   bestS =PRA(S1, S2, objectiveFunctionS)
   return bestS
}

PRA(S1, S2, objectiveFunction)
{
   bestS = better from S1 and S2 due to objectiveFunction
   S = S1
   for(pi = 0; pi < n; pi++)
   {
      pj = in S find position of activity from position pi in S2
      makeMove(S, pj, pi, m_pj)
      for(all modes of found activity)
         S = best schedule due to objectiveFunction
      if(S is better than bestS due to objectiveFunction) bestS = S
   }
   return bestS
}
```

Fig. 5 Pseudo-codes of the TA_PRA and PRA algorithms

individual could be repeated several times. If this does not produce enough feasible solutions the infeasible ones are added to the population in the common memory. In some instances the initial population consist of the infeasible solutions only. Individuals in the common memory are represented as (M,S). The final solution is obtained from the schedule by the procedure based on SGSU.

The time and frequency an agent of each kind receives a solution or set of solutions from the common memory with a view to improve its quality is determined by the strategy. For solving the MRCPSP/max problem instances the strategy with blocking has been used where individuals forwarded to optimization agents for improvement are randomly chosen from the population stored in the common memory. Such individuals are sent to optimization agents ready to start searching for a better solution. After computation the improved individual replaces the one which was send. Additionally, if some solutions (in this approach 5) are received but the best solution in the population has not been improved a new one is generated randomly. It replaces the worst one in the population.

5 Computational Experiment

To validate the proposed approach and to evaluate the effectiveness of the optimization agents the computational experiment has been carried out using benchmark instances of MRCPSP/max from PSPLIB [19], [15], [16] - test set mm100 with activities carried out in 3, 4 and 5 modes. The set includes 270 problem instances. The experiment involved computation with the fixed number of optimization agents representing TA_LSAm, TA_LSAe, TA_TSAe, TA_CA, and TA_PRA algorithms, fixed population size, and the limited time period allowed for computation. Values are chosen on the basis of the previous experiments [1], [13], [14].

The discussed results have been obtained using 5 optimization agents - one of each kind. Population of solutions in the common memory consisted of 5, 10, 15, 20, 25 and 30 individuals. The computation has been stopped after 1, 2 or 3 minutes (Stop time) if no better solution is found. The optimization algorithms have had the fixed and randomly chosen parameter values. For example, in the case of LSAm and LSAe algorithms the *startActivityPosition* has been chosen randomly from the interval $[1, step]$. The *iterationNumber* and *step* parameters in the LSAm and LSAe algorithms has been fixed to 30 and 1 if the *objectiveFunctionN* and *objectiveFunctionF* have been used and to 30 and 2 if the *objectiveFunctionS* has been used. The same parameters in the TSAe algorithm has been fixed to 10 and 1 if the *objectiveFunctionN* and *objectiveFunctionF* have been used and to 10 and 5 if the *objectiveFunctionS* has been used. In the case of CA the *step* parameter has been set to 1 or 5 respectively.

Experiment has been carried out using nodes of the cluster Holk of the Tricity Academic Computer Network built of 256 Intel Itanium 2 Dual Core 1.4 GHz with 12 MB L3 cache processors and with Mellanox InfiniBand interconnections with 10Gb/s bandwidth. During the computation one node per five to eight agents was used.

6 Computational Experiment Results

During the experiment the following characteristics have been calculated and recorded: mean relative error (Mean RE) calculated as the deviation from the lower bound (LB), percent of feasible solutions (% FS), mean computation time required to find the best solution (Mean CT) and mean total computation time (Mean total CT). Each instance has been solved five times and the results have been averaged over these solutions. The results depending on population size are presented in Table 1 and Figure 6. The results depending on population size and number of modes in activities are presented in Table 2 and Figure 7. The computation times and total computation times related to these results are presented in Tables 3, 4 and 5.

Table 1 Performance characteristics (Mean RE) of the proposed A-Team depending on the population size for benchmark test set mm100

Stop time [min]	Population size 5	10	15	20	25	30
1	48.82%	56.22%	81.27%	103.84%	126.29%	118.81%
2	40.61%	41.79%	44.53%	57.40%	76.94%	82.59%
3	35.84%	35.60%	37.84%	44.14%	51.12%	56.05%

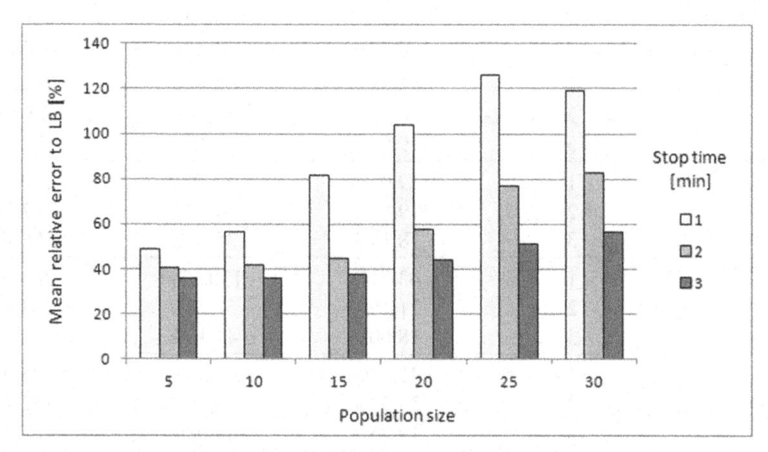

Fig. 6 The graphical representation of the results presented in Table 1

Solutions obtained by the proposed agent-based approach are compared with the results reported in the literature obtained by tabu search algorithm TS_{DR} and heuristic based on multipass priority-rule method with backplanning Prio proposed in [9], as well as the results obtained by double genetic algorithm proposed in [3]. The literature reported results are presented in Table 6.

Experiment results show that the proposed JABAT based A-Team for MRCPSP/max implementation using triple-action agents is effective. The obtained results are better than in the previous implementations [13], [14] where double-action agents are used. The feasible solutions are found for 100% of instances. Presented results are comparable with solutions known from the literature (see Tables 1 and 6).

The results show that the proposed approach is more effective using small populations of results - less than 10 individuals (Figure 6) and is rather time consuming (Tables 3 and 4). Considering population sizes, the best results have been obtained for the longest computations time - 3 minutes, in all cases. Considering computation times, the best results has been obtained for the smallest number of individuals - 5, in all cases. It may be caused by two reasons: (1) in the proposed approach the population management is not effective enough or (2) there are too many infeasible solutions in a bigger populations resulting in a relatively long period needed to produce a critical mass of feasible solutions. Probably both these factors influence the computation times.

Table 2 Performance characteristics (Mean RE) of the proposed A-Team depending on the population size and the number of modes in one activity for benchmark test set mm100

Population size	Stop time [min]	Modes number 3	4	5
5	1	24.28%	41.79%	80.39%
	2	21.99%	39.24%	60.60%
	3	20.33%	34.66%	52.53%
10	1	23.62%	49.22%	95.83%
	2	21.94%	38.79%	64.65%
	3	20.89%	34.71%	51.21%
15	1	25.58%	68.93%	149.30%
	2	20.26%	37.80%	75.53%
	3	19.04%	35.10%	59.37%
20	1	29.05%	101.77%	180.70%
	2	19.87%	44.62%	107.70%
	3	18.19%	35.83%	77.68%
25	1	43.77%	130.17%	204.92%
	2	21.75%	65.43%	143.64%
	3	18.60%	41.16%	93.59%
30	1	42.72%	120.67%	193.02%
	2	23.73%	71.58%	152.44%
	3	19.55%	45.08%	103.51%

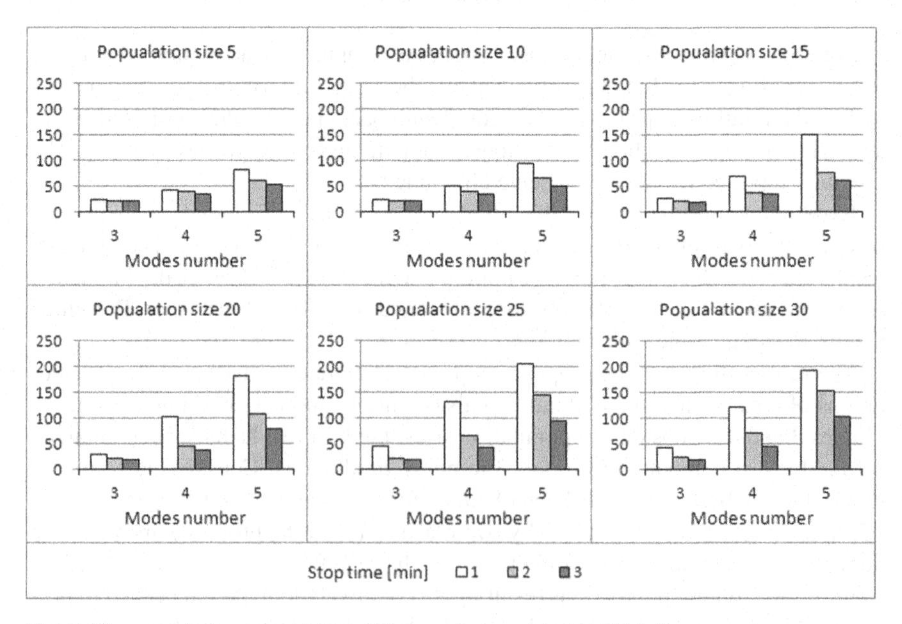

Fig. 7 The graphical representation of the results presented in Table 2

Table 3 Mean computation time in seconds for the proposed A-Team for benchmark test set mm100

Stop time [min]	Population size 5	10	15	20	25	30
1	164.6	163.5	122.3	83.0	56.2	56.1
2	242.0	264.9	319.7	284.5	220.0	195.5
3	270.4	300.3	512.1	439.9	425.6	387.2

Table 4 Mean total computation time in seconds for the proposed A-Team for benchmark test set mm100

Stop time [min]	Population size 5	10	15	20	25	30
1	195.5	194.1	152.7	112.9	86.9	86.5
2	303.4	325.8	380.5	344.8	280.3	255.6
3	362.2	391.4	603.0	530.1	516.9	477.4

Table 5 Mean computation time and total computation time in seconds needed by the proposed A-Team depending on the population size and the number of modes in one activity for benchmark test set mm100

Population size	Stop time [min]	Modes number 3	4	5
5	1	118.2/149.1	189.7/220.6	185.9/216.9
	2	140.7/201.9	257.7/319.2	327.7/389.0
	3	144.5/236.1	258.3/350.1	408.4/500.4
10	1	138.0/169.4	188.6/219.4	163.7/193.6
	2	152.7/214.0	281.1/341.9	361.0/421.4
	3	159.9/250.3	289.3/381.1	451.7/542.7
15	1	164.2/195.2	137.9/168.3	64.9/94.7
	2	224.5/285.3	367.1/427.8	367.5/428.5
	3	311.7/401.7	539.7/631.7	684.9/775.6
20	1	148.1/178.4	66.8/96.8	34.2/63.5
	2	276.7/338.0	355.7/416.3	221.2/280.0
	3	319.2/409.7	528.9/618.9	471.6/561.6
25	1	106.0/136.6	36.6/67.5	25.9/56.8
	2	298.7/359.4	255.1/315.5	106.0/165.9
	3	361.9/453.4	536.6/628.1	430.7/521.8
30	1	101.4/131.8	41.2/71.2	25.6/56.4
	2	291.7/351.1	206.2/266.0	88.6/149.6
	3	390.1/480.5	489.4/579.5	282.2/372.2

Table 6 Literature reported results for benchmark test set mm100

#Modes	Mean RE	% FS	Mean CT [s]
		Literature reported results – TS_{DR} [9]	
3	40%	53%	100
4	91%	61%	100
5	164%	67%	100
		Literature reported results – Prio [9]	
3	63%	100%	100
4	113%	100%	100
5	170%	100%	100
		Literature reported results – DGA [3]	
3,4 and 5	22.05%	100%	100

A serious disadvantage of the proposed approach is the long computation time. The data construction method used in the A-Team for MRCPSP/max implementation could be possibly improved. However, the second factor which is the communication overhead is quite hard to reduce in case of the multi-agent systems. Holding a big number of solutions returned by optimization agents are important for the results quality. It can be however observed that more often agents return solutions the message queues in the system become longer. The bottleneck of the proposed system is the SolutionManager implementation.

Considering the results with respect to the number of modes (Figure 7) it can be observed that the proposed approach is more effective for projects consisting of activities with less number of modes. In the case of 5 modes the differences between results for different stop times are far more noticeable as in the case of 3 or 4 modes. Additionally, these differences are grater for a bigger populations. It could be probably related to the nature of algorithms proposed to solve the MRCPSP/max problem instances.

Apart from the performance evaluation measured in terms of mean relative error defined as deviation from the lower bound and computation times the influence of each agent performance was evaluated. The observed measure was the average percent of individuals which were improved by an agent and the average percent of the current best solutions found by it. Population sizes of 5, 10 ad 15 instances are considered. The results calculated as the average percent of solutions improved and the best improved by each agent and percent of non-improved solutions are presented in Figure 8 and Tables 7 and 8.

It can be observed (Figure 8) that in the proposed approach more effective agents are LSAe, LSAm and TSAe and less effective ones are CA and PRA. The effectiveness of all agents increases with an increase of the computation time. The highest number of solutions have been received from LSAm agent which was the quickest one. Unfortunately it has produced quite a big number of non improved solutions. The LSAe and TSAe produce significantly less solutions than LSAm but they

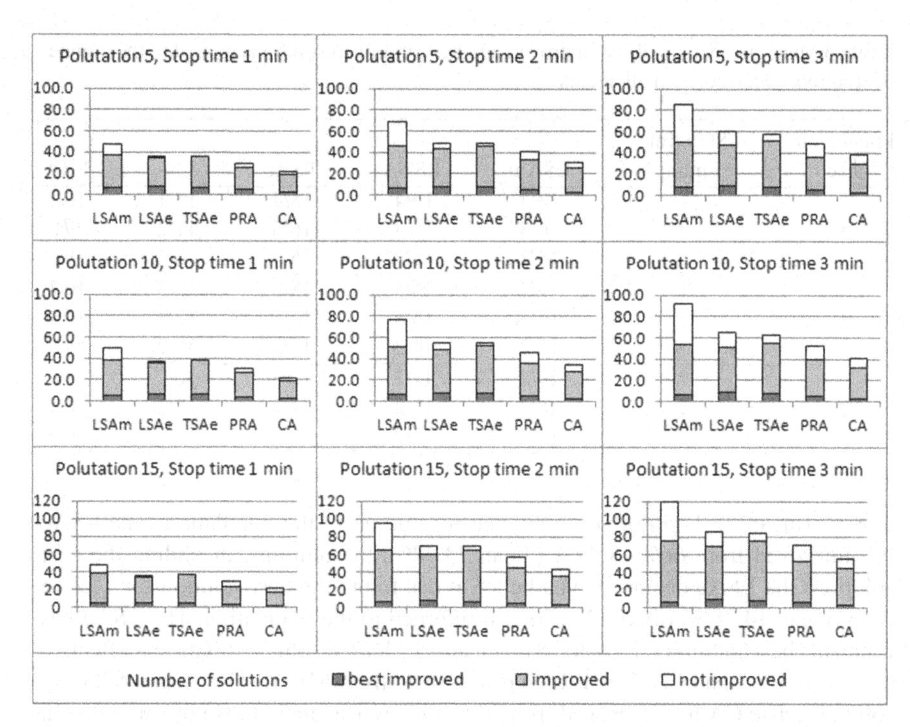

Fig. 8 Mean numbers of the best improved, improved and non improved solutions produced by each agent

Table 7 Mean percentage of the improved and best improved solutions related to the number of solutions received from each agent

Population size	Stop time [min]	LSAm	LSAe	TSAe	PRA	CA
	1	65/13	79/21	85/18	73/16	76/10
5	2	57/10	72/16	81/15	67/13	73/8
	3	51/8	65/14	76/14	61/12	70/7
	1	67/12	80/19	86/16	73/14	77/9
10	2	59/9	73/15	82/13	68/12	75/8
	3	52/8	66/13	76/12	63/11	72/7
	1	71/9	83/14	89/12	67/10	72/7
15	2	62/6	75/11	84/10	68/9	76/7
	3	57/6	71/10	80/9	66/9	75/6

improve or best improve almost all solutions that they have received. The lowest number of solutions produced the CA agent. The number of the best improved and improved solutions returned by CA is the lowest as well.

Considering effectiveness of agents in terms of average percents of improved and best improved solutions returned to the population by each agent (Table 7), also most effective ones are the TSAe and LSAe. The TSAe has improved from 76%

Table 8 Mean percentage of the improved and best improved solutions related to the number of solutions received from all agents

Population size	Stop time [min]	LSAm	LSAe	TSAe	CA	PRA
	1	18/4	16/4	18/4	13/3	10/1
5	2	16/3	15/3	16/3	11/2	9/1
	3	15/2	14/3	15/3	10/2	9/1
	1	19/3	16/4	18/3	13/3	10/1
10	2	17/3	15/3	17/3	12/3	10/1
	3	15/2	14/3	15/2	11/2	9/2
	1	20/2	17/3	19/3	12/2	9/1
15	2	18/2	15/2	17/2	12/2	10/1
	3	16/2	15/2	16/2	11/1	10/1

to 89% (mean 82.11%) of solutions that it returned to the population. The LSAe has improved from 66% to 83% (mean 73.78%) of solutions. In finding the best solutions the LSAe occurred even better, it has found the best solution for 10% to 21% (mean 14.78%) of solutions that it returned to the population. The TSAe has found the best solution for 9% to 18% (mean 13.11%) of solutions. The less effective agents are PRA and CA. The PRA has improved from 61% to 73% (mean 67.33%) solutions. The CA has improved from 70% to 77% (mean 74.00%) of solutions. In finding the best solutions the CA is the less effective, it has found the best solution only for 6% to 10% (mean 7.67%) of solutions. The PRA has found the best solution for 8% to 14% (mean 11.22%) of solutions.

Evaluating particular agents using the criterion of the numbers of returned improved and best improved solutions in total (Table 8), it can be noted that their contributions towards improving the current best solution in the population is quite similar. It varies from 9% to 20% (mean 14.07%) of the improved solutions and from 1% to 4% (mean 2.33%) of the best improved solutions. From this point of view the LSAm and TSAe have occurred the most effective agents and again PRA and CA have been the less effective ones.

7 Conclusions

Experiment results show that the proposed E-JABAT based A-Team implementation is an effective tool for solving instances of the MRCPSP/max problem. Presented results are comparable with solutions known from the literature and better than in the previous implementation. The main disadvantage of the proposed approach is rather long computation time.

Future research can concentrate on improving the implementation details in order to shorten the computation time. Also, the role and implementation of the Solution-Manager should be considered. Dividing the SolutionManager's tasks into more agents should effected on significantly better computation times. The other part of research can focus on finding the best configuration of the heterogeneous agents and parameter settings for their algorithms. It is interesting which agents should or should not be replicated to improve the results. Additionally, testing and adding to E-JABAT other different optimization agents and improving the existing ones could be considered. The other possibility is finding and testing different or additional objective functions for MRCPSP/max problem which could be used in the algorithms.

Acknowledgements. The research has been supported by the Ministry of Science and Higher Education grant no. N N519 576438 for years 2010–2013. Calculations have been performed in the Academic Computer Centre TASK in Gdansk.

References

1. Barbucha, D., Czarnowski, I., Jędrzejowicz, P., Ratajczak-Ropel, E., Wierzbowska, I.: E-JABAT - An Implementation of the Web-Based A-Team. In: Nguyen, N.T., Jain, L.C. (eds.) Intelligent Agents in the Evolution of Web and Applications, pp. 57–86. Springer, Heidelberg (2009)
2. Bartusch, M., Mohring, R.H., Radermacher, F.J.: Scheduling Project Networks with Resource Constraints and Time Windows. Annual Operational Research 16, 201–240 (1988)
3. Barrios, A., Ballestin, F., Valls, V.: A double genetic algorithm for the MRCPSP/max. Computers and Operations Research 38, 33–43 (2011)
4. Glover, F.: Tabu search and adaptive memory programing: Advances, applications and challenges. In: Barr, R.S., Helgason, R.V., Kennington, J.L. (eds.) Interfaces in Computer Scinece and Operations Research, pp. 1–75. Kluwer (1996)
5. Glover, F., Laguna, M.: Tabu Search. Kluwer Academic Publishers (1997)
6. Glover, F.: Tabu search - Part I. ORSA Journal on Computing 1, 190–206 (1989)
7. Glover, F.: Tabu search - Part II. ORSA Journal on Computing 2, 4–32 (1989)
8. Glover, F., Laguna, M., Marti, R.: Fundamentals of scatter search and path relinking. Control and Cybernetics 39, 653–684 (2000)
9. Heilmann, R.: Resource-Constrained Project Scheduling: a Heuristic for the Multi-Mode Case. OR Spektrum 23, 335–357 (2001)
10. Heilmann, R.: A branch-and-bound procedure for the multi-mode resource-constrained project scheduling problem with minimum and maximum time lags. European Journal of Operational Research 144, 348–365 (2003)
11. Jędrzejowicz, P., Wierzbowska, I.: JADE-Based A-Team Environment. In: Alexandrov, V.N., van Albada, G.D., Sloot, P.M.A., Dongarra, J. (eds.) ICCS 2006, Part III. LNCS, vol. 3993, pp. 719–726. Springer, Heidelberg (2006)
12. Jędrzejowicz, P., Ratajczak-Ropel, E.: Solving the RCPSP/max Problem by the Team of Agents. In: Håkansson, A., Nguyen, N.T., Hartung, R.L., Howlett, R.J., Jain, L.C. (eds.) KES-AMSTA 2009. LNCS (LNAI), vol. 5559, pp. 734–743. Springer, Heidelberg (2009)
13. Jędrzejowicz, P., Ratajczak-Ropel, E.: A-Team for Solving MRCPSP/max Problem. In: O'Shea, J., Nguyen, N.T., Crockett, K., Howlett, R.J., Jain, L.C. (eds.) KES-AMSTA 2011. LNCS (LNAI), vol. 6682, pp. 466–475. Springer, Heidelberg (2011)
14. Jędrzejowicz, P., Ratajczak-Ropel, E.: Double-Action Agents Solving the MRCPSP/Max Problem. In: Jędrzejowicz, P., Nguyen, N.T., Hoang, K. (eds.) ICCCI 2011, Part II. LNCS, vol. 6923, pp. 311–321. Springer, Heidelberg (2011)

15. Kolisch, R., Sprecher, A., Drexl, A.: Characterization and generation of a general class of resource constrained project scheduling problems. Management Science 41, 1693–1703 (1995)
16. Kolisch, R., Sprecher, A.: PSPLIB-a project scheduling problem library. European Journal of Operational Research 96, 205–216 (1997)
17. Neumann, K., Schwindt, C., Zimmermann, J.: Project Scheduling with Time Windows and Scarce Resources, 2nd edn. Springer, Heidelberg (2003)
18. Nonobe, K., Ibaraki, T.: A Tabu Search Algorithm for a Generalized Resource Constrained Project Scheduling Problem. In: MIC 2003: The Fifth Metaheuristics International Conference, pp. 55-1–55-6 (2003)
19. PSPLIB, `http://129.187.106.231/psplib`
20. Reyck, B., Herroelen, W.: The Multi-Mode Resource-Constrained Project Scheduling Problem with Generalized Precedence Relations. European Journal of Operational Research 119, 538–556 (1999)
21. Talukdar, S., Baerentzen, L., Gove, A., de Souza, P.: Asynchronous Teams: Co-operation Schemes for Autonomous, Computer-Based Agents. Technical Report EDRC 18-59-96. Carnegie Mellon University, Pittsburgh (1996)

Team of A-Teams - A Study of the Cooperation between Program Agents Solving Difficult Optimization Problems

Dariusz Barbucha, Ireneusz Czarnowski, Piotr Jędrzejowicz,
Ewa Ratajczak-Ropel, and Izabela Wierzbowska*

Abstract. The chapter investigates effects and impact of cooperation between the co-operating A-Teams working in parallel and combined into an architecture de-signed for solving difficult combinatorial optimization problems. Computational experiments carried-out using the available benchmark datasets have confirmed that architectures enabling some kind of cooperation may be competitive in terms of the quality of solutions in comparison with architectures that use traditional, non-cooperating, teams of agents. Also, it has been shown that results may im-prove when cooperating teams are heterogenous, e.g. each consists of different types of agents.

1 Introduction

As it has been observed in [2] the techniques used to solve difficult combinatorial optimization problems have evolved from constructive algorithms to local search techniques, and finally to population-based algorithms.

Since the publication of Goldberg seminal work [11] different classes of evolutionary algorithms have been developed including genetic algorithms, genetic pro-gramming, evolution strategies, differential evolution, cultural evolution, coevolu-tion and population learning algorithms. Not much later studies of the social be-havior of organisms have resulted in development of swarm intelligence systems including ant colony optimization and particle swarm optimization.

In recent years, technological advances have enabled development of various parallel and distributed versions of the population based methods. At the same time, as a result of convergence of many technologies within computer science such as object-oriented programming, distributed computing and artificial life, the agent

Dariusz Barbucha · Ireneusz Czarnowski · Piotr Jędrzejowicz ·
Ewa Ratajczak-Ropel · Izabela Wierzbowska
Department of Information Systems, Gdynia Maritime University,
Morska 83, 81-225 Gdynia, Poland
e-mail: {barbucha, irek, pj, ewra, iza}@am.gdynia.pl

* Corresponding author.

I. Czarnowski et al. (Eds.): Agent-Based Optimization, SCI 456, pp. 123–141.
DOI: 10.1007/978-3-642-34097-0_6

technology has emerged. An agent is understood here as any piece of soft-ware that is designed to use intelligence to automatically carry out an assigned task, mainly retrieving, processing and delivering information.

Paradigms of the population-based methods and multiple agent systems have been during mid nineties integrated within the concept of the asynchronous team of agents (A-Team). A-Team is a multi agent architecture, which has been pro-posed in several papers of S.N. Talukdar and co-authors [18], [19], [20], [21].

Acording to Talukdar [21] an asynchronous team is a collection of software agents that cooperate to solve a problem by dynamically evolving a population of solutions. As Rachlin et al. [17] observed agents cooperate by sharing access to populations of candidate solutions. Each agent works to create, modify or remove solutions from a population. The quality of the solutions gradually evolves over time as improved solutions are added and poor solutions are removed. Cooperation between agents emerges as one agent works on the solutions produced by another. Within an A-Team, agents are autonomous and asynchronous. Each agent encapsulates a particular problem-solving method along with the methods to decide when to work, what to work on and how often to work.

A-Team architecture could be classified as a software multi-agent system that is used to create software assistant agents. According to Baerentzen [1] an asynchronous team (A-Team) is a network of agents (workers) and memories (repositories for the results of work). The paper claims that it is possible to design A-Teams to be effective in solving difficult computational problems. The main design issues are structure of the network and the complement of agents.

The middleware platforms supporting implementation of A-Teams are represented by the JADE-Based A-Team environment (JABAT). Its subsequent versions and extensions were proposed in [3], [9] and [12]. The JABAT middleware was built with the use of JADE (Java Agent Development Framework), a software framework proposed by TILAB [6]. JABAT complies with the requirements of the next generation A-Teams which are portable, scalable and in conformity with the FIPA standards. To solve a single task (i.e. a single problem instance) JABAT uses a population of solutions that are improved by optimizing agents which represent different optimization algorithms. In traditional A-Teams agents work in parallel and independently and cooperate only indirectly using a common memory containing population of solutions.

In [14] JABAT environment has been extended through integrating the team of asynchronous agent paradigm with the island-based genetic algorithm concept first introduced in [8]. A communication, that is information exchange, between cooperating A-Teams has been introduced. It has been shown that in experiments with the Euclidean planar travelling salesman problem (EPTSP) a noticeable improvement in the quality of the computation results has been achieved. In this chapter more problems are examined.

The chapter extends earlier research reported in [4] and is constructed as follows: Section 2 describes concept of *TA-Teams* which is an implementation of a set of specialized and cooperating A-Teams designed to solve instances of difficult combinatorial optimization problems. Section 3 gives details of the experiment settings and a discussion of the experiment results with a view to compare models with or without communication between respective Teams. Section 4 contains description and results of the experiment carried out to assess models with the heterogenous and homogenous A-Teams. Finally, in Section 5, some conclusions and suggestions for future research are drawn.

2 Team of A-Teams - Concept, Implementation and Settings

JABAT middleware environment can be used to implement A-Teams that run in parallel and produce solutions to optimization problems using a set of optimizing agents, each representing an improvement algorithm. Such an algorithm receives a solution and attempts to improve it. Afterwards, successful or not, the result is send back to where it came from. The process of solving a single task (that is an instance of the problem at hand) consists of several steps. At first the initial population of solutions is generated and stored in the common memory. Individuals forming the initial population are, at the following computation stages, improved by independently acting agents (called optimization agents), each executing an improvement algorithm, usually problem dependent (Fig. 1). Different improvement algorithms executed by different agents supposedly increase chances for reaching the global optimum. After a number of reading, improving and storing back cycles, when the stopping criterion is met, the best solution in the population is taken as the final result.

A-Team architecture offers several advantages. Its main advantage is ability to produce good quality solutions to difficult optimization problems. Rachlin et al. [17] mention modularity, suitability for distributed environments and robustness. It is also clear that A-Teams demonstrate properties of the collective intelligence system since the collective of agents can produce better solutions than individual members of such collective.

A-Teams are based on a starlike topology. According to Zhu [26], in a starlike topology the activities of the agents are coordinated or administered by some supervisory (or facilitator) agents designated in the assembly. Only agents that have connections built and specified to the coordinator can interact with each other. An advantage of starlike topology is its loosely enforced control and coordination. Though control and coordination limits the boundary of cooperation the agents can reach, it is desirable when efficiency of cooperation is a main issue that needs to be ensured. According to Cheyer and Martin [7] the use of facilitators (in our case *Task Managers* or *Solution Managers*) offers both advantages and weaknesses with respect to scalability and fault tolerance. For example, on the plus side, the grouping of a facilitator with a collection of client agents provides a natural building block from which to construct larger systems. On the minus side, there is the potential for a facilitator to become a communication bottleneck, or a critical point of failure.

A typical JABAT implementation allows for running a number of A-Teams in parallel providing the required computational resources are available, however the teams never communicate and produce results independently. The implementation of Teams of A-Teams allows for a number of A-Teams to solve the same task in

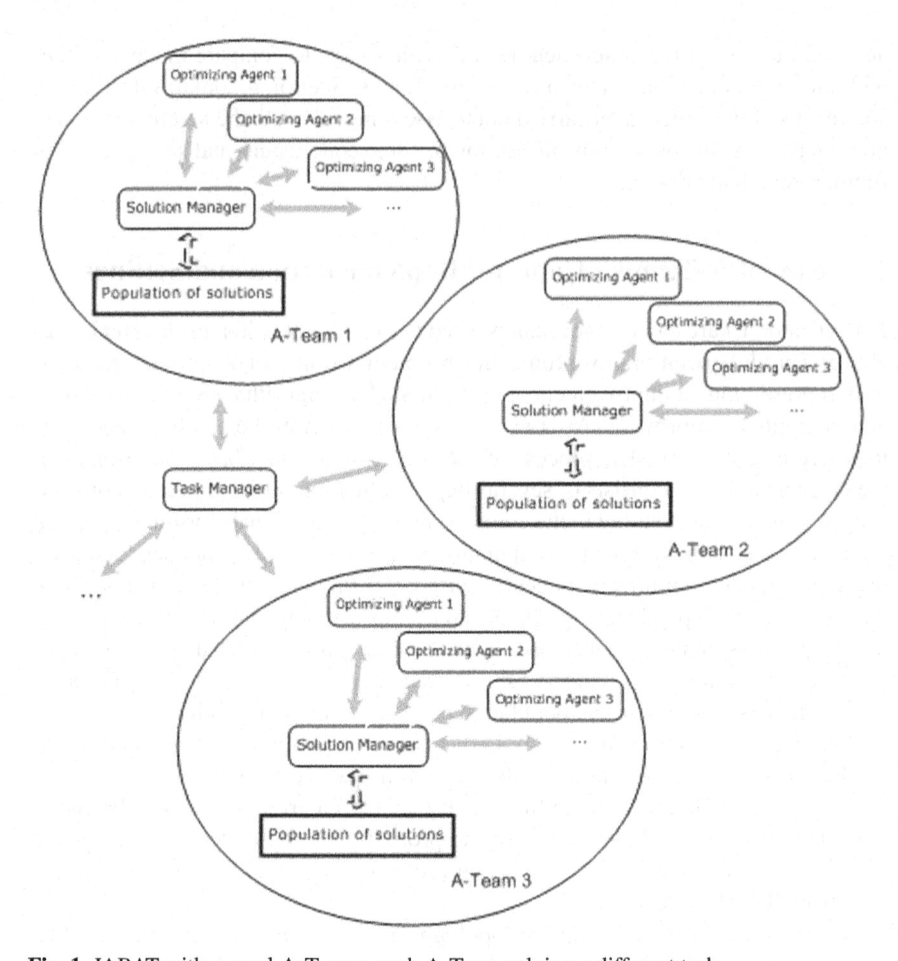

Fig. 1 JABAT with several A-Teams, each A-Team solving a different task

parallel by exploring different regions of the search space, with the added process of communication, that makes it possible to exchange some solutions between common memories maintained by each of the A-Teams with a view to prevent premature convergence and assure diversity of individuals. Similar idea of carrying out the evolutionary process within subpopulations before migrating some individuals to other islands and then continuing the process in cycles involving evolutionary processes and migrations was previously used in, for example, [22] or [25].

In *TA-Teams* JABAT implementation of the process of communication between common memories is supervised by a specialized agent called *Migration Manager* (Fig. 2) and defined by a number of parameters including:

- **Migration size** - number of individuals sent between common memories of A-Teams in a single cycle
- **Migration frequency** - length of time between migrations

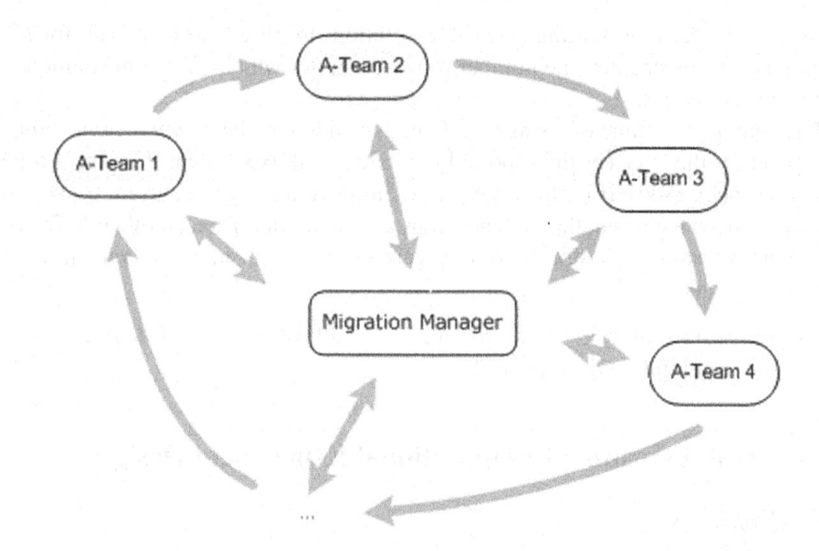

Fig. 2 Team of A-Teams - communication through cyclically exchanging some solutions

- **Migration topology** - an architecture in which an A-Team receives communication from another A-Team and sends communication to some other A-Team. For the purpose of this chapter one-way ring was applied as the chosen topology.
- **Migration policy** - a rule determining how the received solution is incorporated into a common memory of the receiving A-Team.

2.1 Working Strategy

The process of solving a single task in JABAT by an A-Team is controlled by the agent called *Solution Manager* that executes and supervises the, so called, working strategy understood as a set of rules applicable to managing and maintaining the common memory. Common memory contains a population of solutions called individuals. All individuals are feasible solutions of the particular problem instance to be solved.

A-Teams in *TA-Teams* follow the working strategy known as RB-RE which before was identified as the approach assuring generation of a good quality solutions [2]. In this strategy:

- All individuals in the initial population of solutions are generated randomly.
- Selection of individuals for improvement is a random move, however once selected individual (or individuals) can not be selected again until all other individuals have been tried.
- Returning individual replaces the first found worse individual. If a worse individual can not be found within a certain number of reviews (where review is understood as a search for the worse individual after an improved solution is returned) then the worst individual in the common memory is replaced by the randomly

generated one, representing a feasible solution. In all experiments run for the purpose of this chapter, the number of reviews after which a random solution is generated equals 5.

- The computation time of a single A-Team is defined by the *no improvement time gap* set by the user (in the reported computational experiment 2 minutes time gap has been used). If in this time gap no improvement of the current best solution has been achieved, the A-Team stops computations. Then all other A-Teams solving the same task stop as well, regardless of recent improvements in their best solutions.

The overall best result from common memories of all A-Teams in *TA-Teams* is taken as the final solution found for the task.

3 Team of A-Teams - Computational Experiment Design

3.1 Problems

The experiment aimed at investigating effects of agents' cooperation within the $TA-Team$. It has been decided to implement the $TA-Team$ designed to solve difficult computational problems through applying the distributed population-based paradigm. Cooperation effects have been assessed by comparing results obtained from solving benchmark instances of several well-known combinatorial optimization problems by the proposed *TA-Teams* with the results obtained without cooperation between $TA-Team$ members and the results obtained by a single A-Team.

The following combinatorial optimization problems have been selected to be a part of the experiment:

- Euclidean planar traveling salesman problem (EPTSP),
- Vehicle routing problem (VRP),
- Clustering problem (CP),
- Resource-constrained project scheduling problem (RCPSP).

The Euclidean planar traveling salesman problem is a particular case of the general TSP. Given n cities (points) in the plane and their Euclidean distances, the problem is to find the shortest TSP-tour, i.e. a closed path visiting each of the n cities exactly once.

The vehicle routing problem can be stated as the problem of determining optimal routes through a given set of locations (customers) and defined on a directed graph $G=(V,E)$, where $V=\{0,1,\ldots,N\}$ is the set of nodes and E is the set of edges. Node 0 is a depot with NV identical vehicles of capacity W. Each other node $i \in V-\{0\}$ denotes a customer with a non-negative demand d_i. Each link $(i,j) \in E$ denotes the shortest path from customer i to j and is described by the cost c_{ij} of travel from i to j $(i,j=1,\ldots,N)$. It is assumed that $c_{ij}=c_{ji}$. The goal is to find vehicle routes which minimize total cost of travel (or travel distance) such that each route starts and ends at the depot, each customer is serviced exactly once by a single vehicle and the total load on any vehicle associated with a given route does not exceed vehicle capacity.

The clustering problem can be defined as follows. Given a set of N data objects, partition the data set into K clusters, such that similar objects are grouped together and objects with different features belong to different groups. Clustering arbitrary data into clusters of similar items presents the difficulty of deciding what similarity criterion should be used to obtain a good clustering. It can be shown that there is no absolute "best" criterion which would be independent of the final aim of the clustering. Euclidean distance and squared Euclidean distance are probably the most commonly chosen measures of similarity. Partition is understood as providing for each data object an index or label of the cluster to which it is assigned. The goal is to find such a partition that minimizes the objective function, which, in our case, is the sum of squared distances of the data objects to their cluster representatives.

The resource-constrained project scheduling problem consists of a set of n activities, where each activity has to be processed without interruption to complete the project. The dummy activities 1 and n represent the beginning and the end of the project. The duration of the activity $j, j = 1, \ldots, n$ is denoted by d_j where $d_1 = d_n = 0$. There are r renewable resource types. The availability of each resource type k in each time period is r_k units, $k = 1, \ldots, r$. Each activity j requires r_{jk} units of resource k during each period of its duration where $r_{1k} = r_{nk} = 0, k = 1, \ldots, r$. All parameters are non-negative integers. There are precedence relations of the finish-start type (FS) with a zero parameter value (i.e. $FS = 0$) defined between the activities. In other words, activity i precedes activity j if j cannot start until i has been completed. The structure of the project can be represented by an activity-on-node network $G = (SV, SA)$, where SV is the set of activities and SA is the set of precedence relationships. SS_j (SP_j) is the set of successors (predecessors) of activity $j, j = 1, \ldots, n$. It is further assumed that $1 \in SP_j, j = 2, ..., n$, and $n \in SS_j, j = 1, \ldots, n-1$. The objective is to find a schedule S of activities starting times $[s_1, \ldots, s_n]$, where $s_1 = 0$ and resource constraints are satisfied, such that the schedule duration $T(S) = s_n$ is minimized. The above formulated problem is a generalization of the classical job shop scheduling problem.

3.2 Optimizing Agents

To solve instances of each of the above described problems four specialized *TA-Teams* have been designed and implemented:

Instances of the EPTSP are solved with the use of the following optimization algorithms:

- *Simple exchange* (Ex2) - deletes two random edges from the input solution thus breaking the tour under improvement into two disconnected paths and reconnects them in the other possible way, reversing one of them.
- *Recombination*1 (R1) - there are two input solutions. A subpath from one of them is randomly selected. In the next step it is supplemented with edges from the second solution. If this happens to be impossible to add an edge, as the node has already been used in the subpath, the procedure constructs an edge connecting endpoint of the subpath with the closest point in the second input solution not yet in the resulting tour.

- *Mutation* (M) - two randomly selected points from the input solution are directly connected. This subpath is supplemented with edges from the input solution, as in R1.

Instances of the VRP are solved with the use of the following optimization algorithms:

- *Opti3Opt* - an agent which is an implementation of the 3-opt local search algorithm, in which for all routes first three edges are removed and next remaining edges are reconnected in all possible ways.
- *Opti2Lambda* - an implementation of the local search algorithm based on λ- interchange local optimization method [13]. It operates on two selected routes and is based on the interchange/move of customers between routes. For each pair of routes from an individual a parts of routes of length less than or equal to λ are chosen and next these parts are shifted or exchanged, according to the selected operator. Possible operators are defined as pairs: (v,u), where $u,v = 1,\ldots,\lambda$ and denote the lengths of the part of routes which are moved or exchanged. For example, operator $(2,0)$ indicates shifting two customers from the first route to the second route, operator (2,2) indicates an exchange of two customers between routes. Typically, $\lambda = 2$ and such value was used in the reported implementation.
- *OptiLSA* - an implementation of local search algorithm which operates on two selected routes. First, a node (customer) situated relatively far from the centroid of the first route is removed from it and next it is inserted to the second route.

Instances of the CP are solved with the use of the following optimization procedures:

- *OptiRLS* (Random Local Search) - a simple local search algorithm which finds the new solution by exchanging two randomly selected objects belonging to different clusters. The new solution replaces the current one if it is an improvement.
- *OptiHCLS* (Hill-Climbing Local Search) - a simple local search algorithm which finds the new solution by allocating the randomly selected object, from the randomly chosen cluster, to the cluster with minimal euclidean distance to the mean vector. The new solution replaces the current one if it is an improvement.
- *OptiTS* (Tabu Search) - a local search algorithm with tabu active list, which modifies the current solution by allocating the randomly selected object not on the tabu list from the randomly chosen cluster, to other randomly selected cluster. Next, the object is placed on the tabu list and remains there for a given number of iterations. The new solution replaces the current one if it is an improvement.

Instances of the RCPSP are solved with the use of the following optimization procedures:

- *OptiLSA* (Local Search Algorithm) - an implementation of simple local search algorithm which finds local optimum by moving each activity to all possible places in the schedule. For each combination of activities the value of possible solution is calculated. The best schedule is returned.
- *OptiCA* (Crossing Algorithm) - an algorithm based on using the one point crossover operator. Two initial solutions are repeatedly crossed until a better

Table 1 Instances used in the reported experiment

Problem	Instance name and dimensions	Source
EPTSP	pr76 (76 cities), pr144, (144) pr299 (299), pr439 (439), pr1002 (1002)	TSPLIB [23]
VRP	vrpnc1 (50 customers), vrpnc2 (75), vrpnc3 (100), vrpnc4 (150), vrpnc5 (199)	OR-Library [15]
CP	Ruspini (75 objects, 2 attributes), Cleveland heart disease (303, 13), Credit approval (690,15), Iris (150, 4)	UCI Machine Learning Repository [24]
RCPSP	j3013_01 (30 activities), j3029_09 (30), j6009_02 (60), j9021_03 (90), j12016_01 (120)	PSPLIB [16]

solution will be found or all crossing points will be checked. The frequency of crossing is determined by the argument.

- *OptiPRA* (Path-Relinking Algorithm) - an implementation of the path-relinking algorithm. For a pair of solutions a path between them is constructed. The path consists of schedules obtained by carrying out a single move from the preceding schedule. The move is understood as moving one of the activities to a new position. For each schedule in the path the value of the respective solution is checked. The best schedule is remembered and finally returned.

In the computational experiment each of the four *TA-Teams* implementations has been used to solve several instances of the respective combinatorial optimization problems. All these instances have been taken from several well-known benchmark datasets libraries, as shown in Table 1.

3.3 Architectures

Each instance chosen for the experiment has been solved with the use of three different architectures:

- A1 - single A-Team,
- A2 - 5 A-Teams without communication nor migration feature,
- A3 - fully functional *TA-Teams* (with migration) consisting of 5 A-Teams.

Table 2 Experiment settings

	problem	EPTSP		VRP, CP		RCPSP	
	architecture	A1	A2, A3	A1	A2, A3	A1	A2, A3
number of common memories		1	5	1	5	1	5
number of individuals	in one common memory	35	7	50	10	250	50
	in total		35		50		250
number of optimizing agents of each kind	cooperating with one common memory	5	1	5	1	5	1
	in total						5

For the communication within the A3 architecture the following parameter settings have been used:

- *Migration size* = 1 (in one cycle one individual is sent from the common memory of an A-Team to the common memory of another A-Team)
- *Migration frequency* = 0.3 minute
- *Migration topology* - we consider a one-way ring architecture, in which each A-Team receives communication from one adjacent A-Team and sends communication to another adjacent A-Team.
- *Migration policy* - *best-worst* policy, in which the best solution taken from the source population replaces the worst solution in the target population.

The other settings are shown in Table 2 and guarantee that for each problem the complexity in all three architectures remains similar: the total number of individual solutions and agents is the same.

Experiment has been carried out on the cluster Holk of the Tricity Academic Computer Network built of 256 Intel Itanium 2 Dual Core with 12 MB L3 cache processors with Mellanox InfiniBand interconnections with 10Gb/s bandwidth.

As it has been mentioned before *TA-Teams* have been implemented using JABAT middleware derived from JADE. As a consequence it has been possible to create agent containers on different machines and connecting them to the main platform. Then agents may migrate from the main platform to these containers. Each instance used in the reported experiment was solved with the use of 5 nodes of the cluster - one for the main platform and four for the optimising agents to migrate to.

For each problem and each architecture there were no less then 30 runs. Computation errors have been calculated in relation to the best results known for the instances of the investigated problems. Finally, the results - in terms of relative computation error - have been averaged.

3.4 Computational Experiment Results

Table 3 shows mean values and standard percentage deviations of the respective fitness functions calculated for all computational experiment runs as specified in the experiment design described in the previous section.

The main question addressed by the reported experiment is to decide whether the choice of the architecture has influence on the quality of solutions obtained by

Table 3 Mean values and standard percentage deviations of the respective fitness functions calculated for all computational experiment runs

Instance	Architecture					
	A1		A2		A3	
pr76	108904,8	0,8%	108628,1	0,5%	108382,1	0,4%
pr144	58714,8	0,4%	58636,4	0,2%	58606,1	0,2%
pr299	49046,7	0,9%	49481,0	0,8%	48862,7	0,7%
pr439	111860,8	1,5%	111209,5	1,2%	110381,6	1,5%
pr1002	274172,0	1,5%	279592,8	0,6%	273698,2	1,1%
vrpnc1	524,6	0,0%	524,6	0,0%	524,6	0,0%
vrpnc2	851,8	2,0%	853,9	2,2%	851,0	1,9%
vrpnc3	837,1	1,3%	837,8	1,4%	837,8	1,4%
vrpnc4	1064,8	3,5%	1067,5	3,8%	1062,8	3,3%
vrpnc5	1376,0	6,5%	1379,4	6,8%	1367,5	5,9%
Austr. Credit	1923,7	8,7%	1718,4	3,7%	1655,0	3,4%
Cleveland Heart	951,8	14,5%	888,5	4,0%	862,0	3,3%
Iris2	295,4	25,0%	270,3	13,6%	227,3	10,5%
Iris3	157,4	19,5%	141,7	13,0%	121,4	16,5%
Iris4	167,5	9,3%	133,5	17,9%	134,2	17,8%
Ruspini2	3549,8	5,8%	3406,8	2,3%	3369,3	2,4%
Ruspini3	2342,5	7,2%	2317,7	11,4%	2083,0	15,1%
Ruspini4	1752,2	5,9%	1841,6	12,9%	2017,6	18,8%
j30_29_9	98,4	1,0%	98,1	0,9%	98,3	0,9%
j60_9_2	85,2	0,5%	84,7	0,4%	84,9	0,4%
j60_17_4	71,0	0,0%	71,0	0,0%	71,0	0,0%
j90_21_3	129,4	0,6%	128,7	0,6%	128,6	0,6%
j120_16_1	213,0	0,6%	212,3	0,6%	212,3	0,6%

A-Teams? To answer this question the one-way analysis of variance (ANOVA) for each of the considered instances has been carried out. For the purpose of the analysis the following hypotheses have been formulated:

- H0 - zero hypothesis: the choice of architecture does not influence the quality of solutions (mean values of the fitness function are not statistically different).
- H1 - alternative hypothesis: the quality of solutions is not independent from the architecture used (mean values of the fitness function are statistically different).

The analysis has been carried out at the significance level of 0.05. The results are shown in Table 4 and allow to observe the following:

- Zero hypothesis stipulating that the choice of architecture does not influence the quality of solutions (mean values of the fitness function are not statistically different) is accepted in 8 out of 20 considered cases. However, in 2 cases out of 8 (e.g. vrpnc1 and j60_17_4) all three architectures gave exactly the same - optimal - solutions.

Table 4 ANOVA test results for considered problems

Problem	Instance name and dimensions	Source
CP	CreditTrening	-
	HeartTrening	-
	Iris2	yes
	Iris3	-
	Iris4	yes
	Ruspini2	yes
	Ruspini3	-
	Ruspini4	yes
EPTSP	pr76	-
	pr144	yes
	pr299	-
	pr439	-
	pr1002	-
VRP	vrpnc1	yes
	vrpnc2	-
	vrpnc3	-
	vrpnc4	-
	vrpnc5	-
RCPSP	j30_29_9	yes
	j60_17_4	yes
	j60_9_2	-
	j90_21_3	-
	j120_16_1	-

- In 12 out of 20 cases zero hypothesis is rejected in favour of the alternative hypothesis suggesting that the quality of solutions is not independent from the architecture used (mean values of the fitness function are statistically different).

Taking into account that the problems under consideration are not homogenous it has not been possible to carry out analysis of variance with reference to suitability of particular architecture to particular problem. Instead it has been decided to use the non-parametric Friedman test in order to obtain the answer to the question whether particular architectures are equally effective independently of the kind of problem being solved.

The above test has been based on weights (points) assigned to architectures used in the experiment: 1, 2 or 3 points for the worse, second worse and best architecture. The test aimed at deciding among the following hypotheses:

- H0 - zero hypothesis: considered architectures are statistically equally effective regardless of the kind of problem,
- H1 - alternative hypothesis: not all architectures are equally effective.

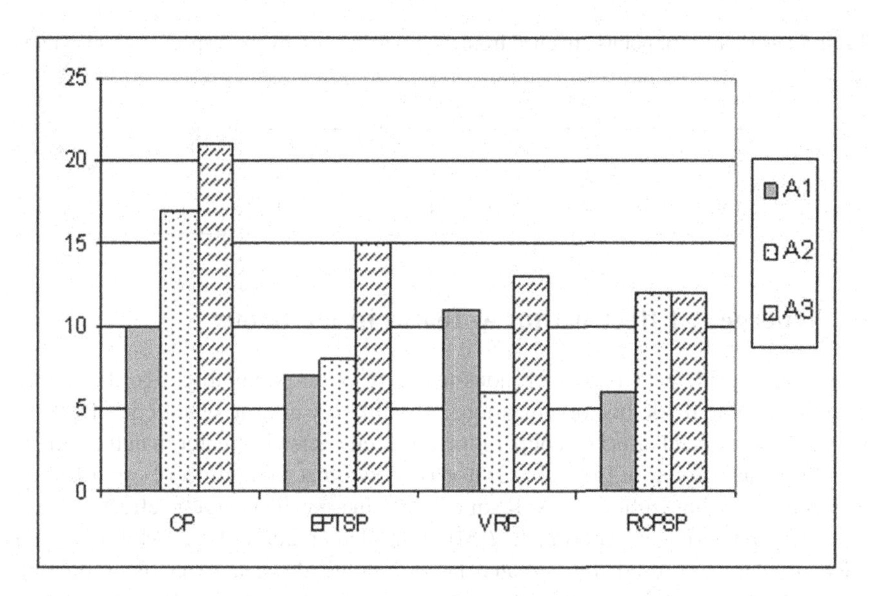

Fig. 3 The Friedman test weights for each problem and each architecture

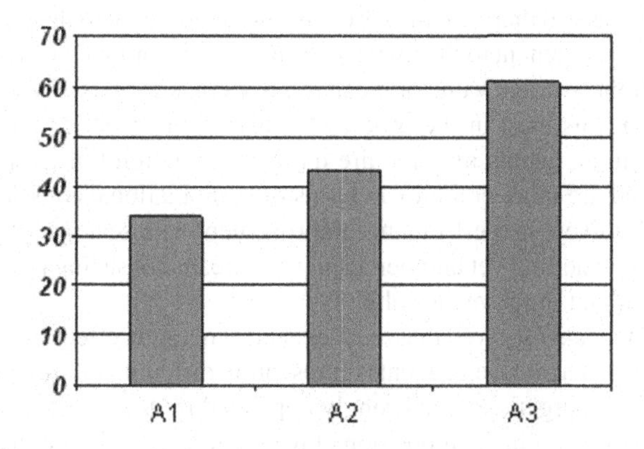

Fig. 4 The total of Friedman weights for the investigated architectures

The analysis has been carried out at the significance level of 0.05. The respective value of the χ^2 statistics with 3 architectures and 23 instances of the considered problems is equal to 18 and the value of χ^2 distribution is equal to 5.99. Thus it can be observed that not all architectures are equally effective regardless of the kind of problem of which instances are being solved.

In Figures 3 and 4 sums of weights obtained by the architectures for each of the considered problems and overall total of weights for each architecture are, respectively, compared. The highest score in total was obtained by architecture A3. Within the problems A3 also scored best except of RCPSP problem, where A1 and A2 both scored the same total. Besides, in Table 5 distribution of points obtained from the Friedman test by respective architectures is shown.

Table 5 Distribution of points obtained from the Friedman test by the respective architectures

Points	A1	A2	A3
1	13	7	1
2	8	11	7
3	2	4	13

4 Synergetic Effect within A-Teams in *TA-Teams*

The ground principal of asynchronous teams rests on combining algorithms into effective problem solving organizations, possibly creating a synergetic effect, in which the combined effect of cooperation between agents is greater than the sum of their separate efforts. In [5] and [13] it was shown that joint effort of optimization agents cooperating within one A-Team may produce such synergetic effect.

In the first series of experiments ([5]) an implementation of JABAT for solving VRP problem was used. Investigated factors included the number of optimizing agents and composition of the team. Two criteria were evaluated: mean relative error and computation time.

It could be observed that in many cases adding a new agent to the team of running agents brings improvement to the final solution. Indeed, for many instances tested in the experiment the average value of mean relative error decreases when the number of optimizing agents used in the process of solving the problem increases. However, applying more agents may require more computation time, especially when the agents added are more complex in terms of computational complexity. What is more, scale-effectiveness (performance improvement with scale) could be observed only up to a point; adding yet another agent to the team consisting of a certain number of agents did not improve the solution.

The experiment also showed that there existed synergetic effect from cooperation of different types of agents, e.g. agents representing different optimizing algorithms. The strength of synergetic effect strongly depended on the structure of the agents team and often diversifying composition of the team resulted in producing solutions of a better quality.

Another experiment was described in [13]. In this case an implementation of JABAT for solving EPTSP was used, with 5 types of relatively low quality optimizing agents. The experiment involved solving EPTSP tasks with the use of different sizes of the common population of solutions. The number of individuals in the population ranged from 2 to 15. A synergetic effect appeared, on average, for all sizes of the common memory.

For a larger problem instances such an effect to appear required some computation time lag. However, small size instances could be effectively solved using a number of homogenous local-search optimizing agents working together for a certain time, which would be shorter than the time required by a heterogeneous A-Team to achieve a comparable quality. All these findings refer to the working strategy used in the experiment where escaping from the local optima was based on inserting to the common memory some random solutions from time to time during the computation process. Unfortunately the experiment results do not allow to draw any general rules for construction of an A-Team from heterogeneous optimizing agents with a view to achieving a synergetic effect, even for the specific problem type.

The existence of the synergetic effect was also shown by the experiment described in [10]. There, tasks from data reduction problem were solved with the use of two different agent teams. There were two homogenous teams consisting of agents implementing simulated annealing and tabu search algorithms respectively, and eight teams consisting of both types of agents (simulated annealing and tabu search algorithms), varying by some parameters of the algorithms. It was shown that the results obtained from heterogenous teams were more accurate. In fact each heterogenous team gave better results than both homogenous teams.

It was decided to check whether the synergetic effect of optimization agents cooperating within each A-Team in the fully functional *TA-Teams* (with migration) still may be observed. To check the existence and strength of such an effect additional experiment has been conducted for EPTSP problem.

4.1 Synergetic Effect in TA-Teams *- Computational Experiment*

The experiment was run for EPTSP problem. The optimising agents and problem instances were the same as described in Section 3. The architectures were chosen as follows:

- B1 - 6 heterogenous A-Teams without communication nor migration feature,
- B2 - fully functional *TA-Teams* (with migration) consisting of 6 homogenous A-Teams,
- B3 - fully functional *TA-Teams* (with migration) consisting of 6 heterogenous A-Teams.

In each heterogenous A-Team there are three different optimising agents: one Ex2, one R1 and one M. In each homogenous A-Team only agents of the same kind cooperate: in B2 there are two A-Teams, each consisting of three Ex2 agents, two A-Teams with three R1 agents, and two with three M agents.

Other settings are the same as in the experiment from Section 3. Again, the settings guarantee that the complexity in all three architectures remains similar: in each architecture the total number of individual solutions and agents is the same.

Also, the experiment was run in the same environment, on the cluster Holk with five nodes, four of which were used for the optimizing agents to migrate to.

For each problem and each architecture there were no less then 30 runs. Computation errors have been calculated in relation to the best results known for the problems. The results - in terms of relative computation error - have been averaged.

4.2 Computational Experiment Results

Table 6 shows mean values and standard percentage deviations of the respective fitness functions calculated for all computational experiment runs as specified in the experiment design described in the previous subsection.

Table 6 Mean values and standard percentage deviations of the respective fitness functions calculated for all computational experiment runs

Problem	B1		B2		B3	
pr76	108539,5	0,4%	108748,9	0,5%	108396,3	0,4%
pr144	58636,0	0,1%	58669,5	0,2%	58578,0	0,1%
pr299	49480,9	0,7%	49322,0	1,1%	48874,7	0,7%
pr439	111391,2	1,1%	111750,0	1,9%	109922,6	1,5%
pr1002	284967,5	1,4%	281580,4	1,5%	272905,0	0,7%

Again, the question addressed by the reported experiment is to decide whether the choice of the architecture has influence on the quality of solutions obtained by A-Teams? To answer this question the one-way analysis of variance (ANOVA) for each of the considered problems has been carried out. For the purpose of the analysis the following hypotheses have been formulated:

- H0 - zero hypothesis: the choice of architecture does not influence the quality of solutions (mean values of the fitness function are not statistically different).
- H1 - alternative hypothesis: the quality of solutions is not independent from the architecture used (mean values of the fitness function are statistically different).

The analysis has been carried out at the significance level of 0.05. From Table 7 we can see that for all considered instances of EPTSP problem the choice of architecture influences the results.

Table 7 ANOVA test results for considered problems

Problem	H0 accepted
pr76	-
pr144	-
pr299	-
pr439	-
pr1002	-

Additionally, to identify differences between architectures, the Tukey test has been carried out, based on the ANOVA results. This test aimed at comparing pairs of the architectures with a view to evaluate which pairs are statistically different and which are not. Architectures that do not statistically differ are shown in Table 8.

Table 8 Pairs of the architectures showing no statistical differences in the Tukey test (pairs marked by X)

Problem	B1/B2	B2/B3	B1/B3
pr76	x		x
pr144	x		
pr299	x		
pr439	x		
pr1002			

The following observation may be drawn: architectures B1 and B2 (heterogenous A-Teams without communication and homogenious *TA-Teams*) produce results of the statistically comparable quality, though the results may differ for bigger tasks.

Fig. 5 compares average relative computation errors for all three architectures. It can be seen that for all data considered the best results (or the smallest average error) was produced by architecture B3 - fully functional *TA-Teams*. For bigger instances, pr299 and pr1002, architecture B2 was worse than architecture B1.

Thus, the above considerations for EPTSP problem allow to draw the conclusion that the architecture with heterogenous *TA-Teams* might be recommended as the one with the highest probability of producing the above average quality result.

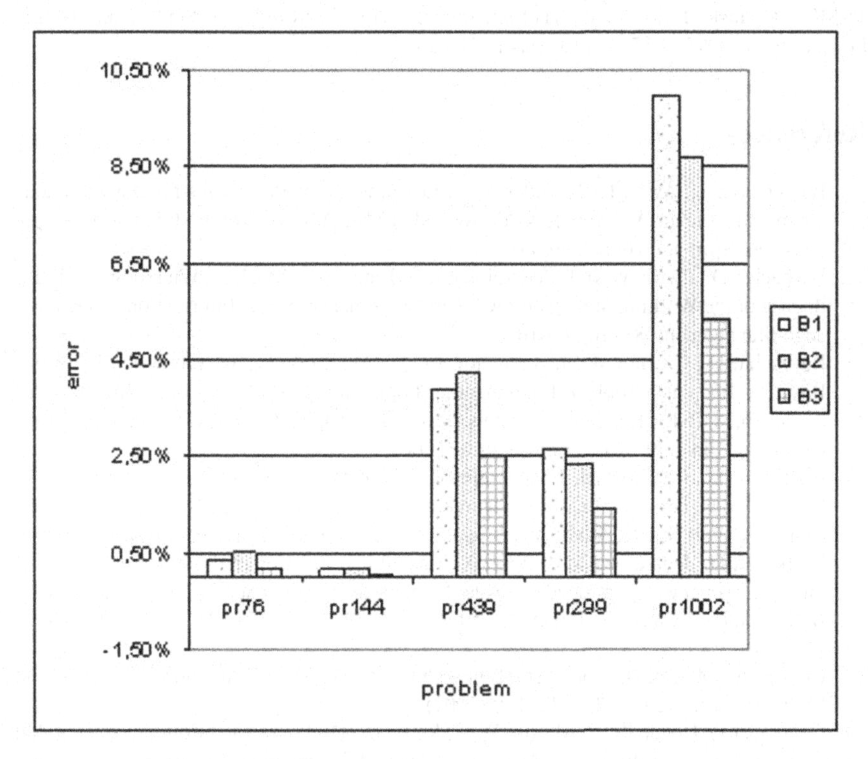

Fig. 5 Average relative computation error

5 Conclusions

The presented research has confirmed that in most cases integrating the distributed evolutionary concept, and especially the island based evolutionary algorithm with the A-Team paradigm might result in achieving a noticeable improvement in the quality of the computation results. Thus, the chapter confirms the importance of choosing an effective architecture.

In [5] and [13] it has been shown that heterogenous A-Team may give better results than homogenious A-Teams. This chapter confirms this results for teams of A-Teams. Thus, heterogeneous A-Teams in the architecture with some kind of cooperation may be competitive in terms of the quality of solutions in comparison with both using traditional, non-cooperating, agents or teams of agents or using only homogenious A-Teams in the same architecture.

The range of parameters used in the experiments is insufficient to draw further conclusions as to strategy of choosing features of the *TA-Teams* constructed from heterogeneous optimizing agents with a view to achieving best results. Future research will focus on evaluating effects of a wider set of parameters and solving more test data, which might be helpful in identifying more general observations.

Acknowledgements. Calculations have been performed in the Academic Computer Centre TASK in Gdansk. The research has been supported by the Ministry of Science and Higher Education no. N N519 576438 for years 2010-2013.

References

1. Baerentzen, L., Avila, P., Talukdar, S.N.: Learning Network Designs for Asynchronous Teams. In: Boman, M., Van de Velde, W. (eds.) MAAMAW 1997. LNCS, vol. 1237, pp. 177–196. Springer, Heidelberg (1997)
2. Barbucha, D., Czarnowski, I., Jędrzejowicz, P., Ratajczak-Ropel, E., Wierzbowska, I.: Influence of the Working Strategy on A-Team Performance. Smart Information and Knowledge Management, 83–102 (2010)
3. Barbucha, D., Czarnowski, I., Jędrzejowicz, P., Ratajczak, E., Wierzbowska, I.: JADE-Based A-Team as a Tool for Implementing Population-Based Algorithms. In: Chen, Y., Abraham, A. (eds.) Intelligent Systems Design and Applications, ISDA, Jinan Shandong China, pp. 144–149. IEEE, Los Alamos (2006)
4. Barbucha, D., Czarnowski, I., Jędrzejowicz, P., Ratajczak-Ropel, E., Wierzbowska, I.: Parallel Cooperating A-Teams. In: Jędrzejowicz, P., Nguyen, N.T., Hoang, K. (eds.) ICCCI 2011, Part II. LNCS, vol. 6923, pp. 322–331. Springer, Heidelberg (2011)
5. Barbucha, D., Jędrzejowicz, P.: An experimental investigation of the synergetic effect of multiple agents working together in the A-Team. System Science 34(2), 55–62 (2008)
6. Bellifemine, F., Caire, G., Poggi, A., Rimassa, G.: JADE. A White Paper Exp. 3(3), 6–20 (2003)
7. Cheyer, A., Martin, D.: The open agent architecture. Journal of Autonomous Agents and Multi-Agent Systems 4(1), 143–148 (2001)
8. Cohoon, J.P., Hegde, S.U., Martin, W.N., Richards, D.: Punctuated Equilibria: a Parallel Genetic Algorithm. In: Proceedings of the Second International Conference on Genetic Algorithms, pp. 148–154. Lawrence Erlbaum Associates, Hillsdale (1987)

9. Czarnowski, I., Jędrzejowicz, P., Wierzbowska, I.: A-Team Middleware on a Cluster. In: Håkansson, A., Nguyen, N.T., Hartung, R.L., Howlett, R.J., Jain, L.C. (eds.) KES-AMSTA 2009. LNCS (LNAI), vol. 5559, pp. 764–772. Springer, Heidelberg (2009)
10. Czarnowski, I., Jędrzejowicz, P.: Agent-based Simulated Annealing and Tabu Search Procedures Applied to Solving the Data Reduction Problem. Int. J. Appl. Math. Comput. Sci. 21(1), 57–68 (2011)
11. Goldberg, D.E.: Genetic Algorithms in Search, Optimization, and Machine Learning. Addison-Wesley, Reading (1989)
12. Jędrzejowicz, P., Wierzbowska, I.: JADE-Based A-Team Environment. In: Alexandrov, V.N., van Albada, G.D., Sloot, P.M.A., Dongarra, J. (eds.) ICCS 2006, Part III. LNCS (LNAI), vol. 3993, pp. 719–726. Springer, Heidelberg (2006)
13. Jędrzejowicz, P., Wierzbowska, I.: Experimental Investigation of the Synergetic Effect Produced by Agents Solving Together Instances of the Euclidean Planar Travelling Salesman Problem. In: Jędrzejowicz, P., Nguyen, N.T., Howlet, R.J., Jain, L.C. (eds.) KES-AMSTA 2010, Part II. LNCS (LNAI), vol. 6071, pp. 160–169. Springer, Heidelberg (2010)
14. Jędrzejowicz, P., Wierzbowska, I.: Parallel Cooperating A-Teams Solving Instances of the Euclidean Planar Travelling Salesman Problem. In: O'Shea, J., Nguyen, N.T., Crockett, K., Howlett, R.J., Jain, L.C., et al. (eds.) KES-AMSTA 2011. LNCS (LNAI), vol. 6682, pp. 456–465. Springer, Heidelberg (2011)
15. OR-Library, `http://people.brunel.ac.uk/mastjjb/jeb/orlib/vrpinfo.html`
16. PSPLIB, `http://129.187.106.231/psplib`
17. Rachlin, J., Goodwin, R., Murthy, S., Akkiraju, R., Wu, F., Kumaran, S., Das, R.: A-Teams: An Agent Architecture for Optimization and Decision-Support. In: Müller, J.P., Rao, A.S., Singh, M.P. (eds.) ATAL 1998. LNCS (LNAI), vol. 1555, pp. 261–276. Springer, Heidelberg (1999)
18. Talukdar, S.N.: Collaboration Rules for Autonomous Software Agents. Decision Support Systems 24, 269–278 (1999)
19. Talukdar, S., Baerentzen, L., Gove, A., de Souza, P.: Asynchronous Teams: Cooperation Schemes for Autonomous Agents. Journal of Heuristics 4(4), 295–232 (1998)
20. Talukdar, S.N., Pyo, S.S., Giras, T.: Asynchronous Procedures for Parallel Processing. IEEE Trans. on PAS, PAS-102(11), 3652–3659 (1983)
21. Talukdar, S.N., de Souza, P., Murthy, S.: Organizations for Computer-Based Agents. Engineering Intelligent Systems 1(2), 56–69 (1993)
22. Tanese, R.: Distributed genetic algorithms. In: Schaffer, J. (ed.) Proceedings of the Third International Conference on Genetic Algorithms, pp. 434–439. Morgan Kaufmann, San Mateo (1989)
23. TSP library, `http://comopt.ifi.uniheidelberg.de/software/TSPLIB95`
24. UCI Machine Learning Repository, `http://archive.ics.uci.edu/ml/`
25. Whitley, D., Rana, S., Heckendorn, R.B.: The Island Model Genetic Algorithm: On Separability, Population Size and Convergence. Journal of Computing and Information Technology 7, 33–47 (1998)
26. Zhu, Q.: Topologies of agents interactions in knowledge intensive multi-agent systems for networked information services. Advanced Engineering Informatics 20, 31–45 (2006)

Distributed Bregman-Distance Algorithms for Min-Max Optimization

Kunal Srivastava, Angelia Nedić*, and Dušan Stipanović

Abstract. We consider a min-max optimization problem over a time-varying network of computational agents, where each agent in the network has its local convex cost function which is a private knowledge of the agent. The agents want to jointly minimize the maximum cost incurred by any agent in the network, while maintaining the privacy of their objective functions. To solve the problem, we consider subgradient algorithms where each agent computes its own estimates of an optimal point based on its own cost function, and it communicates these estimates to its neighbors in the network. The algorithms employ techniques from convex optimization, stochastic approximation and averaging protocols (typically used to ensure a proper information diffusion over a network), which allow time-varying network structure. We discuss two algorithms, one based on exact-penalty approach and the other based on primal-dual Lagrangian approach, where both approaches utilize Bregman-distance functions. We establish convergence of the algorithms (with probability one) for a diminishing step-size, and demonstrate the applicability of the algorithms by considering a power allocation problem in a cellular network.

1 Introduction

This work is motivated by coordinator-free distributed algorithms for optimization problems originating in [38, 5], which have recently seen a resurgence driven by wireless network applications. A canonical problem in coordinator-free distributed setting is the problem of reaching an agreement among the local decision variables in a network of computational agents [39]. Protocols for achieving an agreement employ local averaging which is known to be robust to time-varying graph topology and noisy communication links [18]. These averaging protocols have led to a new

Kunal Srivastava · Angelia Nedić · Dušan Stipanović
University of Illinois, Urbana, IL 61801, USA
e-mail: {kkunal2,angelia,dusan}@illinois.edu

* Corresponding author.

I. Czarnowski et al. (Eds.): Agent-Based Optimization, SCI 456, pp. 143–174.
DOI: 10.1007/978-3-642-34097-0_7

class of distributed algorithms for parameter estimation [37], distributed optimization [29, 25, 23, 19, 21, 35] and control of multi-agent systems [16]. Recent work in agreement-based distributed optimization has focused on minimizing the sum of agents' local cost functions [25, 23, 35, 17, 19, 20, 32], which arises in wide variety of areas ranging from sensor networks [28] to distributed machine learning [1].

Optimizing the sum of local objective functions is a popular choice for problems arising in resource allocation and network utility maximization [34], as the objective function is a measure of fair resource allocation. An alternative notion of fairness is the min-max criterion [8], where the interest is in determining an optimal decision variable that minimizes the worst case loss incurred by any agent. However, the algorithms for solving distributed min-max problems over networks are in their infancy. To fill out this void, we have recently considered a distributed method for solving such a problem in [36], where we provided a subgradient algorithm based on the exact-penalty function approach. Here, we make a further progress in two different directions: (1) we present a distributed exact penalty algorithm that uses Bregman distances [10] as opposed to the standard Euclidean distance considered in [36]; and (2) we provide an alternative distributed primal-dual algorithm that also uses Bregman distances. We establish convergence of both algorithms for a diminishing step-size and time-varying network, under mild conditions on the network connectivity. We allow for stochastic errors in subgradient evaluations, which are assumed to be zero-mean and with uniformly bounded expected norm, as typically done in stochastic approximations or stochastic subgradient methods [13, 14, 27, 6, 9, 26].

The min-max optimization aspect addressed in this paper is novel with respect to the prior work on distributed optimization over a network, which is dealing exclusively with the minimization of the sum of agent objectives [29, 25, 23, 19, 21, 1, 35, 32], except for [36]. The network aspect of this paper sets it apart from the standard optimization problems with noisy (sub)-gradient evaluations [14, 27, 6, 9, 26].

The advantage of the proposed algorithms is their ability to solve a class min-max distributed problems for which currently there are no specific algorithms, except for [36]. The difficulty in developing such algorithms comes from the inherent distributed knowledge of the problem data in a network of agents. Thus, there is a need for an algorithm that has ability to "align" the iterates among the agents while simultaneously solving the network problem. Both of our proposed algorithms achieve this task.

Regarding the benefits of the proposed algorithms, the exact-penalty algorithm is simpler for implementation, but in principle it is a distributed subgradient approach due to the use of non-differentiable penalty functions. The primal-dual approach requires a larger number of variables than the penalty approach, but it preserves the smoothness of the problem (provided that the original agent objective functions f_i are smooth). In the presence of stochastic errors in (sub)gradient evaluations, both algorithms will have overall rate of the order of $1/\sqrt{k}$ in terms of the number of iterations k, which is typical for stochastic approximations [27, 26].

The rest of the chapter is organized as follows. In Section 2 we state our problem of interest and suitably reformulate it for the development of distributed algorithms. In Section 3 we present our distributed Bregman-distance based algorithm which utilizes the exact penalty function approach, and we establish the convergence of the algorithm. In Section 4 we develop an alternative algorithm which builds on the primal-dual approach of Arrow-Hurwicz-Uzawa [2], and we prove its convergence. This algorithm paves a way to handling the problems in which the network plays a min-max game against an exogenous signal, which is discussed in Section 5. In Section 6 we present an example of min-max power allocation in a cellular network and provide simulation results for both the exact penalty approach and the primal-dual approach. We conclude in Section 7.

Notation: The set of real numbers is denoted by $\mathbb{R}$, while non-negative real numbers are denoted by $\mathbb{R}_+$. All vectors are viewed as columns, where the jth component of a vector x is denoted by x_j. We use the symbol $\langle x, y\rangle$ to denote the inner-product between two vectors x and y. We write $\mathbf{1}$ for the vector with each component equal to 1. A vector π is *stochastic* if $\pi_i \geq 0$ for all i and $\sum_i \pi_i = 1$.

For an $m \times m$ matrix A, we use A_{ij} or $[A]_{ij}$ to denote its entry in the ith row and j^{th} column. An $m \times m$ matrix W is *stochastic* if $W_{ij} \geq 0$ for all i, j, and $W\mathbf{1} = \mathbf{1}$. A stochastic matrix W is *doubly stochastic* if it satisfies $\mathbf{1}^T W = \mathbf{1}$. Given a directed graph $G = (V, E)$, the link $(i, j) \in E$ is to be interpreted as the incoming edge from j to i. For a bidirectional graph G, we have $(i, j) \in E$ if and only if $(j, i) \in E$. We will sometimes denote the edge set of a graph G as $\mathscr{E}(G)$. We use the terms "agent" and "node" interchangeably. We say that agent j is a *neighbor* of agent i if $(i, j) \in E$, and we denote the set of all neighbors of agent i by N_i. When the edge set is time varying, we use $N_i(t)$ to denote the neighbors of agent i at time t. Given a logical statement $p(x)$ that is predicated on a variable x, we use $\mathbf{1}_{\{p(x)\}}$ to denote the indicator function which takes value 1 when $p(x)$ is $true$ and value 0 when $p(x)$ is $false$.

2 Problem Formulation

We consider a system of m computational agents, which is viewed as the node set $V = \{1, \ldots, m\}$. We assume that the time is discrete and use $k = 0, 1, \ldots$ to denote the time instances. The agents communicate with each other over a time-varying communication network. At any time k, the communications among the agents are represented by a directed graph $G(k) = (V, E(k))$ with an edge-set $E(k)$ that has a link $(i, j) \in E(k)$ if and only if agent i receives information from agent j at time k.

Let each agent i have a cost function f_i, which is known only to that agent. Consider a distributed multi-agent optimization problem subject to local agent communications, where the agents want to cooperatively solve the following problem:

$$\min_{x \in X} \max_{i \in V} f_i(x). \tag{1}$$

Here, each $f_i : \mathbb{R}^n \to \mathbb{R}$ is a *convex function*, representing a local objective function known only by agent i. The set $X \subseteq \mathbb{R}^n$ is a *closed and convex set* known by all agents. Throughout the paper, *we assume that the problem has a nonempty optimal set*, which is denoted by X^*.

We assume that $\mathbb{R}^n$ is equipped with some norm $\|\cdot\|$, with a corresponding dual norm $\|\cdot\|_*$. The goal is to develop a *distributed algorithm* for solving the constrained optimization problem in (1), while obeying the network connectivity structure and local information exchange among the neighboring agents.

The min-max-problem in (1) is a convex problem, as the function $f(x) = \max_{i\in V} f_i(x)$ is convex since point-wise maximum of convex functions preserves convexity ([4], Proposition 1.2.4, page 30). We are interested in the case when the agents' objective functions f_i are not necessarily differentiable. We also allow the local objective functions f_i to take the form of the following stochastic optimization:

$$f_i(x) = \mathbb{E}_{\omega_i}[F_i(x,\omega_i)] + \Omega_i(x), \tag{2}$$

where the expectation is taken with respect to the distribution of a random variable ω_i. The term $\Omega_i(x)$ is a regularization term that may be included to improve the generalization ability [15], or to enforce sparsity on solutions.

Min-max problem (1) does not lend itself to distributed computations that obey the local connectivity of the agents in the network. We find it useful to use the epigraph representation of the problem. In particular, we let $\eta \in \mathbb{R}$ and we re-cast problem (1) in an equivalent form:

$$\begin{aligned} &\text{minimize} && \eta \\ &\text{subject to} && f_i(x) \le \eta \quad \text{for all } x \in X,\ \eta \in \mathbb{R}, \text{ and } i \in V, \end{aligned} \tag{3}$$

where x and η are variables. We use η^* to denote the optimal value of problem (3).

To distributedly solve the min-max problem, we provide two algorithms aimed at solving its epigraph formulation in (3). The first algorithm is based on an exact penalty function approach and the second algorithm is based on a primal-dual approach, where both algorithms employ Bregman-distance functions.

Before proceeding with the algorithmic development, we provide some basics of a Bregman-distance function as introduced by Bregman [10] (also, see for example [11]). Let $\mathbb{R}^n$ be equipped with some norm $\|\cdot\|$, whose dual norm is $\|x\|_* = \sup_{\|y\|\le 1}\langle y,x\rangle$. Let $X \subseteq \mathbb{R}^n$ be a convex set and $\omega : X \to \mathbb{R}$ be a differentiable convex function over X. The function ω is *strongly convex* with a parameter $\sigma > 0$ (with respect to the norm $\|\cdot\|$), if it satisfies

$$\omega(y) - [\omega(x) + \langle \nabla\omega(x), y-x\rangle] \ge \frac{\sigma}{2}\|x-y\|^2 \quad \text{for all } x \in X^\circ \text{ and } y \in X,$$

where X° denotes the relative interior of the set X. Alternatively, ω is strongly convex over X with a parameter $\sigma > 0$ if there holds:

$$\langle \nabla\omega(x) - \nabla\omega(y), x-y\rangle \ge \sigma\|x-y\|^2 \quad \text{for all } x,y \in X^\circ.$$

Given a strongly convex function ω, we can define the *Bregman-distance* function $B : X \times X^\circ \to \mathbb{R}_+$ induced by ω:

$$B(y,x) = \omega(y) - [\omega(x) + \langle \nabla\omega(x), y-x\rangle].$$

This function is also referred to as the *prox-function* induced by ω. For the function B, by the convexity of ω we have

$$B(y,x) \geq 0 \qquad \text{for all } y \in X \text{ and } x \in X^\circ.$$

Furthermore, by the strong convexity of ω, for every $x \in X^\circ$, *the function* $B(\cdot,x)$ *is strongly convex over* X *with the same parameter* σ. The Bregman function $B(y,x)$ is used to define a nonlinear projection operator associated with the given set X, also known as the *prox-operator* [26], as follows:

$$P_X(d,x) = \text{argmin}_{z\in X}\left\{\langle d, z-x\rangle + B(z,x)\right\}.$$

3 Exact Penalty Function Approach

We further transform the problem in (3) by penalizing the constraints to obtain the following problem:

$$\min_{x\in X, \eta\in\mathbb{R}} \; \eta + \sum_{i=1}^{m} r_i g_i(x,\eta), \tag{4}$$

where each g_i is a penalty function given by

$$g_i(x,\eta) = \max\{0, f_i(x) - \eta\}$$

and $r_i > 0$ is a penalty parameter for violating the constraint $f_i(x) \leq \eta$. Under certain conditions [3], the solutions of the penalized problem (4) are also the solutions of the constrained problem (3). Specifically, these conditions involve the Lagrangian dual problem associated with problem (3). We introduce the Lagrangian function:

$$L(x,\eta,\mu) = \eta + \sum_{i=1}^{m} \mu_i (f_i(x) - \eta), \tag{5}$$

where $\mu = (\mu_1,\ldots,\mu_m)'$ is the vector of dual variables satisfying $\mu_i \geq 0$ for all $i \in V$. The dual problem is

$$\max_{\mu\geq 0} q(\mu) \qquad \text{with} \quad q(\mu) = \inf_{x\in X, \eta\in\mathbb{R}} L(x,\eta,\mu). \tag{6}$$

It can be verified that the Slater condition is satisfied for problem (3) and, hence, there is no duality gap between the primal problem (3) and its dual (6). Furthermore, the set of dual optimal solutions is nonempty and bounded. The bound for the dual optimal variables can be found by rewriting the Lagrangian function (5) as follows:

$$L(x,\eta,\mu) = \left(1 - \sum_{i=1}^{m} \mu_i\right)\eta + \sum_{i=1}^{m} \mu_i f_i(x).$$

Thus, $\inf_{\eta \in \mathbb{R}} L(x,\eta,\mu) = -\infty$ when $\sum_{i=1}^{m} \mu_i \neq 1$, implying that $q(\mu) = -\infty$ whenever $\sum_{i=1}^{m} \mu_i \neq 1$. Therefore, the domain of the dual function q is the set of multipliers $\mu \geq 0$ such that $\sum_{i=1}^{m} \mu_i = 1$, showing that optimal multipliers μ_i^* must satisfy

$$\sum_{i=1}^{m} \mu_i^* = 1. \tag{7}$$

Hence, according to [3], when the penalty parameters satisfy $r_i > 1$ for all i, then the problems in (4) and (3) are equivalent.

Penalized problem (4) has a form suitable for distributed computations among the agents, as its objective can be written as $\sum_{i=1}^{m}(\eta/m + r_i g_i(x,\eta))$ and the function $\tilde{F}_i(x,\eta) = \eta/m + r_i g_i(x,\eta)$ can be interpreted as an objective function associated with agent i. In this setting, agent i is the only agent that knows the function f_i and, therefore, this agent is the only agent that deals with the dual variable μ_i associated with the constraint $f_i(x) \leq \eta$. Furthermore, observe that there is no need for any coordination of the penalty values r_i among the agents, as each agent just needs to choose its individual penalty $r_i > 1$.

3.1 Equivalence between Epigraph and Penalty Formulations

We establish an important relation between the optimal solutions of the epigraph formulation (3) of the min-max problem and its penalized counterpart (4). The proof of this relation is basically along the lines of the work in [3], but somewhat shorter as it exploits the special structure of the epigraph formulation (3). Furthermore, the relation is important in our further development and it is not readily available.

To prove the result, we use the saddle-point theorem characterizing the optimal solutions of problem (3) and its dual problem (6), as given for example in [4], Proposition 6.2.4, page 360. The theorem is adjusted to the specific form of our Lagrangian function.

Theorem 1. *The pair (z^*,μ^*) with $z^* = (x^*,\eta^*) \in X \times \mathbb{R}$ and $\mu^* \geq 0$ is a saddle-point of the Lagrangian function $L(z,\mu)$ (i.e., primal-dual optimal pair) if and only if the following relation holds:*

$$L(z^*,\mu) \leq L(z^*,\mu^*) \leq L(z,\mu^*) \qquad \text{for all } z = (x,\eta) \in X \times \mathbb{R} \text{ and } \mu \geq 0.$$

Now, we have the following lemma.

Lemma 1. *Let $\eta^* = \min_{x \in X} \max_{i \in V} f_i(x)$ and $r_i > 1$ for all i. Then, for $g_i(x,\eta) = \max\{0, f_i(x) - \eta\}$ we have*

$$\sum_{i=1}^{m} r_i g_i(x,\eta) + \eta \geq \eta^* \qquad \text{for all } x \in X \text{ and } \eta \in \mathbb{R}.$$

Furthermore, the above inequality holds as equality if and only if $\eta = \eta^*$ *and* $x = x^*$ *for an optimal solution* x^* *of the problem* $\min_{x \in X} \max_i f_i(x)$.

Proof. Consider the definition of the Lagrangian in (5). For a given $x \in X$ and η, let us define the dual variables μ_i such that $\mu_i = r_i$ if $f_i(x) - \eta \geq 0$, and $\mu_i = 0$ if $f_i(x) - \eta < 0$, or compactly $\mu_i = r_i \mathbf{1}_{\{f_i(x) \geq \eta\}}$. Then, we have

$$\sum_{i=1}^{m} r_i \max\{f_i(x) - \eta, 0\} + \eta = \sum_{i=1}^{m} \mu_i (f_i(x) - \eta) + \eta = L(z, \mu).$$

Furthermore, we have $L(z^*, \mu^*) = \eta^*$. Thus, we need to prove that $L(z, \mu) - L(z^*, \mu^*) \geq 0$. For this we use Theorem 1, from which we obtain $-L(z^*, \mu^*) \geq -L(z, \mu^*)$, implying

$$\begin{aligned} L(z,\mu) - L(z^*,\mu^*) &\geq L(z,\mu) - L(z,\mu^*) = \sum_{i=1}^{m} (\mu_i - \mu_i^*)(f_i(x) - \eta) \\ &= \sum_{i=1}^{m} (r_i - \mu_i^*) \mathbf{1}_{\{f_i(x) \geq \eta\}} (f_i(x) - \eta) - \sum_{i=1}^{m} \mu_i^* \mathbf{1}_{\{f_i(x) < \eta\}} (f_i(x) - \eta) \geq 0, \end{aligned}$$

where we have used the decomposition $\mu_i^* = \mu_i^* \mathbf{1}_{\{f_i(x) \geq \eta\}} + \mu_i^* \mathbf{1}_{\{f_i(x) < \eta\}}$, and relations $r_i > 1$ and $1 \geq \mu_i^* \geq 0$ for all i (see (7)).

We next show that the preceding inequality holds as equality if and only if $\eta = \eta^*$ and $x = x^*$ for some $x^* \in X^*$. By the definition of min-max solution, if x^* solves the problem then we have $f_i(x^*) \leq \eta^*$ for all i, implying

$$\sum_{i=1}^{m} r_i \max\{f_i(x^*) - \eta^*, 0\} + \eta^* = \eta^*.$$

Thus, we just need to prove the "only if" part. Assume that for some $\bar{x} \in X$ and $\bar{\eta}$,

$$\sum_{i=1}^{m} r_i \max\{f_i(\bar{x}) - \bar{\eta}, 0\} + \bar{\eta} = \eta^*. \tag{8}$$

Since $\sum_{i=1}^{m} r_i \max\{f_i(\bar{x}) - \bar{\eta}, 0\} \geq 0$, it follows $\bar{\eta} \leq \eta^*$. Let us assume that $\bar{\eta} < \eta^*$. Then, for the equality to hold we must have $f_j(\bar{x}) > \bar{\eta}$ for some j. Thus, $f_{i^*}(\bar{x}) > \bar{\eta}$ for $i^* = \mathrm{argmax}_j f_i(\bar{x})$. By $\eta^* = \min_{x \in X} \max_{i \in V} f_i(x)$ we have $f_{i^*}(\bar{x}) \geq \eta^*$ implying $f_{i^*}(\bar{x}) - \bar{\eta} \geq \eta^* - \bar{\eta} > 0$. Since $r_{i^*} > 1$, it follows that $r_{i^*}(f_{i^*}(\bar{x}) - \bar{\eta}) > \eta^* - \bar{\eta}$. Therefore,

$$\sum_{i=1}^{m} r_i \max\{f_i(\bar{x}) - \bar{\eta}, 0\} + \bar{\eta} > \eta^*,$$

which contradicts (8). Hence, we must have $\bar{\eta} = \eta^*$ in (8). This further yields $\sum_{i=1}^{m} r_i \max\{f_i(\bar{x}) - \eta^*, 0\} = 0$, which by $r_i > 0$ implies $f_i(\bar{x}) \leq \eta^*$ for all i, thus showing that $\bar{x}$ is a min-max solution.

3.2 Penalty-Based Algorithm

Here, we present a distributed multi-agent algorithm for solving the penalty reformulation (4) of the min-max problem, where a penalty function $g_i(x,\eta)$ is associated with an agent i and $r_i > 1$ for all $i \in V$. Let x_k^j and η_k^j be the decision variables of agent j at time k, which are agent j estimates of an optimal solution x^* and the optimal value η^* of the problem, respectively. Recall that the agents' communications at time k are represented with a graph $G(k) = (V, E(k))$, where $(i,j) \in E(k)$ if agent i receives estimates $x^j(k)$ and $\eta^j(k)$ from agent j. To capture this information exchange, we let $N_i(k)$ denote the set of neighbors of agent i, i.e., $N_i(k) = \{j \in V \mid (i,j) \in E(k)\}$. Let us introduce a strongly convex function $\omega_x : X \to \mathbb{R}$, with a parameter $\sigma_x > 0$. We assume that $\mathbb{R}$ is equipped with square norm, i.e., $r \mapsto \frac{1}{2}r^2$, and we introduce a scalar function $\omega_\eta : \mathbb{R} \to \mathbb{R}$ that is strongly convex with respect to this norm, with a parameter $\sigma_\eta > 0$. Let us denote the Bregman-distance functions generated by these strongly convex functions by $B_x(\cdot,\cdot)$, and $B_\eta(\cdot,\cdot)$ respectively, i.e.,

$$B_x(y,u) = \omega_x(y) - [\omega_x(u) + \langle \nabla \omega_x(u), y - u \rangle],$$
$$B_\eta(\phi,\zeta) = \omega_\eta(\phi) - [\omega_\eta(\zeta) + \omega_\eta'(\zeta)(\phi - \zeta)],$$

where $\omega_\eta'(\zeta)$ denotes the derivative of ω_η at ζ. Upon receiving the estimates x_k^j and η_k^j from its neighbors, each agent i performs an intermittent adjustment of its estimates as follows:

$$\begin{bmatrix} \tilde{x}_k^i \\ \tilde{\eta}_k^i \end{bmatrix} = \sum_{j \in N_i(k)} w_{ij}(k) \begin{bmatrix} x_k^j \\ \eta_k^j \end{bmatrix}, \tag{9}$$

where $w_{ij}(k) \geq 0$ is a weight that agent i assigns to its neighbor $j \in N_i(k)$.

For a compact representation of relation (9), let $w_{ij}(k) = 0$ for all $j \notin N_i(k)$ and introduce a matrix W_k with entries $w_{ij}(k)$. With this notation, the intermittent adjustment in (9) can be written as follows:

$$\tilde{z}_k^i = \begin{bmatrix} \tilde{x}_k^i \\ \tilde{\eta}_k^i \end{bmatrix} = \sum_{j=1}^{m} [W_k]_{ij} \begin{bmatrix} x_k^j \\ \eta_k^j \end{bmatrix}. \tag{10}$$

After the intermittent adjustment, each agent i takes a step toward minimizing its own penalty function through an adjustment of the following form: for all $i \in V$,

$$x_{k+1}^i = \operatorname{argmin}_{y \in X} \left[\alpha_k r_i \langle \nabla_x g_i(\tilde{z}_k^i) + \varepsilon_k^i, y \rangle + B_x(y, \tilde{x}_k^i)\right],$$
$$\eta_{k+1}^i = \operatorname{argmin}_{s \in \mathbb{R}} \left[\alpha_k \left(\frac{1}{m} + r_i \nabla_\eta g_i(\tilde{z}_k^i)\right) s + B_\eta(s, \tilde{\eta}_k^i)\right], \tag{11}$$

where $r_i > 1$ and $\alpha_k > 0$ is a step size. The notation $\nabla_x g_i(x,\eta)$ denotes a subgradient of g with respect to x, i.e., the term $\nabla f_i(x) \mathbf{1}_{\{f_i(x) \geq \eta\}}$ where we use $\nabla f_i(x)$ to denote a subgradient of f_i at x. Similarly, $\nabla_\eta g_i(x,\eta)$ denotes the partial derivative of g with respect to η i.e., $\nabla_\eta g_i(x,\eta) = -\mathbf{1}_{\{f_i(x)\geq\eta\}}$.

If $\mathbb{R}^n$ is equipped with the Euclidean norm, and the Bregman distance functions are chosen as $\omega_x(y) = \frac{1}{2}\|y\|^2$ and $\omega_\eta(\zeta) = \frac{1}{2}\zeta^2$, then algorithm (10)–(11) reduces to the standard subgradient-projection method:

$$\begin{bmatrix} x_{k+1}^i \\ \eta_{k+1}^i \end{bmatrix} = \Pi_{X\times\mathbb{R}}\left[\tilde{z}_k^i - \alpha_k r_i\left(\nabla g_i(\tilde{z}_k^i) + \begin{bmatrix} \varepsilon_k^i \\ 0 \end{bmatrix}\right)\right],$$

where Π_K stands for the Euclidean projection on a set K.

Let us now take a closer look at the first update relation in (11). This update involves taking a step along an erroneous subgradient of g_i at point $\tilde{z}_k^i$, i.e., the direction $\nabla_x g_i(\tilde{z}_k^i) + \varepsilon_k^i$ where ε_k^i is a subgradient error. The agent i objective function $g_i(x,\eta) = \max\{f_i(x) - \eta, 0\}$ is not differentiable at the point (x,η) where $f_i(x) - \eta = 0$. At such a point, a subgradient of the function g_i at (x,η) exists since each function f_i is assumed to be convex over the entire space ([4], Proposition 4.2.1). A subgradient of g at such a point is given by

$$\nabla g_i(x,\eta) = \begin{bmatrix} \nabla f_i(x) \\ -1 \end{bmatrix} \mathbf{1}_{\{f_i(x)\geq\eta\}}, \tag{12}$$

where $\nabla f_i(x)$ denotes a subgradient of $f(x)$. Thus, the function g_i also has a nonempty subdifferential set at any point (x,η).

The subgradient error ε_k^i in algorithm (10)–(11) is assumed to be stochastic in order to address a general form of the objective function, as in (2), where the subgradient $\nabla f_i(x)$ is not readily available to us. We adopt a standard approach in stochastic optimization by using an unbiased estimate $\nabla f_i(x) + \bar{\varepsilon}_k^i$ of the subgradient, where $\bar{\varepsilon}_k^i$ is a zero mean random variable. Thus, in (11) we have

$$\varepsilon_k^i = \bar{\varepsilon}_k^i \mathbf{1}_{\{f_i(\tilde{x}_k^i)\geq\tilde{\eta}_k^i\}}.$$

The initial points $x_0^i \in X$ and η_0^i, for $i \in V$, may be selected randomly with a distribution independent of any other sources of randomness in the algorithm.

3.3 Assumptions

Our assumptions on the network are the same as, for example, those in [29, 32].

Assumption 1. *For the weight matrices and the communication graphs, we assume the following:*

(a)[Weights Rule] There exists a scalar $0 < \gamma < 1$ *such that* $[W_k]_{ii} \geq \gamma$ *for all i and k, and* $[W_k]_{ij} \geq \gamma$ *whenever* $[W_k]_{ij} > 0$.

(b)[Doubly Stochasticity] The matrix W_k *is doubly stochastic for all k, i.e.,* $W_k\mathbf{1} = \mathbf{1}$, *and* $\mathbf{1}'W_k = \mathbf{1}'$.

(c) [Connectedness] There exists an integer $Q \geq 1$ such that the graph with the vertex set V and the edge set $\cup_{\tau=kQ}^{(k+1)Q-1} E(\tau)$ is strongly connected for every k.

The assumptions ensure that agent's local variables are properly diffused over time-varying communication networks.

Next, we impose the following assumptions on the subgradients $\nabla f_i(x)$ and the errors, where we use $\partial f_i(x)$ to denote the set of all subgradients of f_i at x.

Assumption 2. *Let the following hold:*

(a) The subgradients of each f_i are bounded over the set X, i.e., there is a scalar $C > 0$ such that $\|\nabla f_i(x)\|_* \leq C$ for all $\nabla f_i(x) \in \partial f_i(x)$, all $x \in X$ and all $i \in V$.
(b) The subgradient errors $\bar{\varepsilon}_k^i$ when conditioned on the point $x = \tilde{x}_k^i$ of the subgradient $\nabla f_i(x)$ evaluation are zero mean, i.e., $\mathbb{E}[\bar{\varepsilon}_k^i \mid \tilde{x}_k^i] = 0$ for all $i \in V$ and $k \geq 0$, with probability 1.
(c) There is a scalar $\nu > 0$ such that $\mathbb{E}[\|\bar{\varepsilon}_k^i\|_*^2 \mid \tilde{x}_k^i] \leq \nu^2$ for all $i \in V$ and $k \geq 0$, with probability 1.

Basically, under Assumptions 2-b and 2-c, the iterations $\{x_k^i\}, i \in V$, of the algorithm in (10)–(11) form a Markov process. In what follows, we will use F_k to denote the past iterates of the algorithm (11), i.e.,

$$F_k = \{x_t^i, \eta_t^i, i \in V, t = 0, 1, \ldots, k\} \qquad \text{for } k \geq 0.$$

Note that, given F_k, the iterates $\tilde{x}_k^i$ and $\tilde{\eta}_k^i$ in (10) are deterministic. In view of this, as a consequence of the subgradient norm and subgradient error boundedness (Assumptions 2-a and 2-c), it can be seen that with probability 1,

$$\mathbb{E}\left[\|\nabla f_i(x) + \bar{\varepsilon}_k^i\|_*^2 \mid F_k\right] \leq (C + \nu)^2 \qquad \text{for all } i \in V \text{ and } k \geq 0.$$

Also, as a result of Assumption 2-a, we have

$$\|\nabla_x g_i(\tilde{z}_k^i)\|_* = \|\nabla f_i(\tilde{x}_k^i)\|_* \mathbf{1}_{\{f_i(\tilde{x}_k^i) \geq \tilde{\eta}_k^i\}} \leq C \quad \text{for all } i \in V \text{ and } k \geq 0.$$

This and the zero-mean error assumption (Assumption 2-b) yield

$$\mathbb{E}[\varepsilon_k^i \mid F_k] = \mathbb{E}[\bar{\varepsilon}_k^i \mathbf{1}_{\{f_i(\tilde{x}_k^i) \geq \tilde{\eta}_k^i\}} \mid F_k] = \mathbb{E}\left[\bar{\varepsilon}_k^i \mid F_k\right] \mathbf{1}_{\{f_i(\tilde{x}_k^i) \geq \tilde{\eta}_k^i\}} = 0.$$

Similarly, as a result of Assumption 2-c we have with probability 1,

$$\mathbb{E}[\|\varepsilon_k^i\|_*^2 \mid F_k] \;=\; \mathbb{E}[\|\bar{\varepsilon}_k^i\|_*^2 \mathbf{1}_{\{f_i(\tilde{x}_k^i) \geq \tilde{\eta}_k^i\}} \mid F_k] \;=\; \mathbb{E}\left[\|\bar{\varepsilon}_k^i\|_*^2 \mid F_k\right] \mathbf{1}_{\{f_i(\tilde{x}_k^i) \geq \tilde{\eta}_k^i\}} \;\leq\; \nu^2.$$

This, in turn implies that with probability 1 for all $i \in V$ and $k \geq 0$,

$$\mathbb{E}\left[\|\nabla_x g_i(\tilde{z}_k^i) + \varepsilon_k^i\|_*^2 \mid F_k\right] \leq (C + \nu)^2. \tag{13}$$

Applying Jensen's inequality, we find that

$$\mathbb{E}\left[\|\nabla_x g_i(z_k^i) + \varepsilon_k^i\|_* \mid F_k\right] \leq C + \nu. \tag{14}$$

By the definition, a Bregman function is convex in its first variable. We further make the assumption on the choice of Bregman-distance functions that requires convexity with respect to the second variable. We depend on this assumption when showing the convergence of our algorithm.

Assumption 3. *Both Bregman-distance functions $B_x(y,z)$ and $B_\eta(\phi,\zeta)$ are convex in their second arguments z and ζ, respectively, for every fixed y and ϕ.*

3.4 Algorithm Convergence

In this section we prove the convergence of algorithm (10)–(11). We use techniques from Lyapunov analysis and the following generalization of the supermartingale convergence theorem, which is also known as the Robbins-Siegmund result as it originated in the work of Robbins and Siegmund [33].

Theorem 2. *Let F_t, $t = 0,1,2,\ldots$, be a filtration such that $F_t \subset F_{t+1}$ for $t \geq 0$. Let $\{X_t\}$, $\{Y_t\}$, $\{Z_t\}$ and $\{g_t\}$ be sequences of non-negativerandom variables that are adapted to the filtration F_t. Assume that for each t, we have with probability 1,*

$$\mathbb{E}[Y_{t+1}|F_t] \leq (1+g_t)Y_t - X_t + Z_t,$$

where $\sum_{t=0}^{\infty} Z_t < \infty$ and $\sum_{t=0}^{\infty} g_t < \infty$ with probability 1. Then, with probability 1, $\sum_{t=0}^{\infty} X_t < \infty$ and the sequence Y_t converges to a nonnegative random variable Y.

Our convergence analysis of algorithm (10)–(11) rests on Theorem 2. In order to use this theorem, we establish two main properties of the algorithm showing that the conditions of the theorem are satisfied. We develop these properties in forthcoming Lemmas 4 and 5. In the development, we make use of an alternative representation of the algorithm that relies on transition matrices, defined as follows:

$$\Phi(k,s) = W_k W_{k-1} \cdots W_s \qquad \text{for all } k,s \text{ with } k \geq s \geq 0. \tag{15}$$

We next state a result from [22] (Corollary 1) on the convergence properties of the matrix $\Phi(k,s)$.

Lemma 2. *Let Assumptions 1 hold. Then, we have $\left|[\Phi(k,s)]_{ij} - \frac{1}{m}\right| \leq \theta \beta^{k-s}$ for all $i,j \in V$ and all $k \geq s \geq 0$, with $\theta = \left(1 - \frac{\eta}{4m^2}\right)^{-2}$ and $\beta = \left(1 - \frac{\eta}{4m^2}\right)^{\frac{1}{Q}}$.*

We will also make use of the following result.

Lemma 3. *Let $\{\gamma_k\}$ be a non-negative scalar sequence such that $\sum_k \gamma_k < \infty$. Then, for any β with $0 < \beta < 1$, we have $\sum_{k=0}^{\infty}\left(\sum_{\ell=0}^{k}\beta^{k-\ell}\gamma_\ell\right) < \infty$.*

Proof. Let $\sum_{k=0}^{\infty}\gamma_k < \infty$. For any integer $M \geq 1$, we have $\sum_{k=0}^{M}\left(\sum_{\ell=0}^{k}\beta^{k-\ell}\gamma_\ell\right) = \sum_{\ell=0}^{M}\gamma_\ell\sum_{t=0}^{M-\ell}\beta^t \leq \sum_{\ell=0}^{M}\gamma_\ell\frac{1}{1-\beta}$, implying $\sum_{k=0}^{\infty}\left(\sum_{\ell=0}^{k}\beta^{k-\ell}\gamma_\ell\right) \leq \frac{1}{1-\beta}\sum_{\ell=0}^{\infty}\gamma_\ell < \infty$. □

In our analysis, we use auxiliary points, namely the instantaneous averages of the iterates x_k^i and η_k^i over $i \in V$, defined by

$$\hat{x}_k = \frac{1}{m}\sum_{i=1}^{m}x_k^i, \qquad \hat{\eta}_k = \frac{1}{m}\sum_{i=1}^{m}\eta_k^i \qquad \text{for all } k \geq 0.$$

We next provide an important result for these averages that we use to assert the convergence properties of our algorithm.

Lemma 4. *Let Assumptions 1 and 2 hold, and let $\sum_{k=0}^{\infty}\alpha_k^2 < \infty$. Then, for the iterates of algorithm* (10)–(11) *we have $\sum_{k=0}^{\infty}\alpha_k\|\hat{x}_k - x_k^i\| < \infty$ and $\sum_{k=0}^{\infty}\alpha_k|\hat{\eta}_k - \eta_k^i| < \infty$ for all $i \in V$, with probability 1.*

Proof. Let us denote the noisy subgradient as $\tilde{d}_k^i = \nabla_x g_i(\tilde{z}_k^i) + \varepsilon_k^i$. Applying the optimality condition for (11), we get

$$\langle\alpha_k r_i \tilde{d}_k^i + \nabla\omega_x(x_{k+1}^i) - \nabla\omega_x(\tilde{x}_k^i), y - x_{k+1}^i\rangle \geq 0 \quad \text{for all } y \in X.$$

Since $\tilde{x}_k^i \in X$, by letting $y = \tilde{x}_k^i$ we have

$$\langle\alpha_k r_i \tilde{d}_k^i + \nabla\omega_x(x_{k+1}^i) - \nabla\omega_x(\tilde{x}_k^i), \tilde{x}_k^i - x_{k+1}^i\rangle \geq 0,$$

which implies

$$\alpha_k r_i\langle\tilde{d}_k^i, \tilde{x}_k^i - x_{k+1}^i\rangle \geq \langle\nabla\omega_x(\tilde{x}_k^i) - \nabla\omega_x(x_{k+1}^i), \tilde{x}_k^i - x_{k+1}^i\rangle \geq \sigma_x\|\tilde{x}_k^i - x_{k+1}^i\|^2,$$

where the last inequality follows by the strong convexity of ω_x. Using Hölder's inequality, we obtain

$$\alpha_k r_i\|\tilde{d}_k^i\|_*\|\tilde{x}_k^i - x_{k+1}^i\| \geq \sigma_x\|\tilde{x}_k^i - x_{k+1}^i\|^2.$$

Therefore $\|\tilde{x}_k^i - x_{k+1}^i\| \leq \alpha_k r_i\frac{\|\tilde{d}_k^i\|_*}{\sigma_x}$, and taking the conditional expectation yields

$$\mathbb{E}\left[\|\tilde{x}_k^i - x_{k+1}^i\| \mid F_k\right] \leq \alpha_k r_i\frac{\mathbb{E}\left[\|\tilde{d}_k^i\|_* \mid F_k\right]}{\sigma_x} \leq \alpha_k r_i\frac{C+\nu}{\sigma_x}, \tag{16}$$

where the last inequality follows from (14) under Assumptions 2-a and 2-c.

Let us now write the iterates as follows

$$x_{k+1}^i = \tilde{x}_k^i + e_k^i \qquad \text{with} \quad e_k^i = x_{k+1}^i - \tilde{x}_k^i. \tag{17}$$

By (16) and the iterated expectation rule, for e_k^i we obtain

$$\mathbb{E}[\|e_k^i\|] = \mathbb{E}\left[\mathbb{E}[\|e_k^i\| \mid F_k]\right] \leq \alpha_k r_i \frac{C+\nu}{\sigma_x} \qquad \text{for all } i \text{ and } k. \tag{18}$$

By taking the average of the relations in (17) over $i = 1,\dots,m$ and using the doubly stochastic property of W_k, we can see that for all k,

$$\hat{x}_{k+1} = \hat{x}_k + \frac{1}{m}\sum_{i=1}^{m} e_k^i. \tag{19}$$

Now, we note that by (17) and the definition of $\tilde{x}_k^i$, we have $x_{k+1}^i = \sum_{j=1}^m [W_k]_{ij} x_k^j + e_k^i$. We then recursively use this equation to relate x_{k+1}^i and x_0^j for $j \in V$. We do so by using the matrices $\Phi(k,s)$ defined in (15) to obtain

$$x_{k+1}^i = \sum_{j=1}^{m} [\Phi(k,0)]_{ij} x_0^j + e_k^i + \sum_{\ell=0}^{k-1}\left(\sum_{j=1}^{m} [\Phi(k,\ell+1)]_{ij} e_\ell^j\right). \tag{20}$$

Similarly, we derive the recursive relation for $\hat{x}_{k+1}$ as given in (19), and obtain:

$$\hat{x}_{k+1} = \hat{x}_0 + \frac{1}{m}\sum_{\ell=0}^{k}\sum_{j=1}^{m} e_\ell^j = \frac{1}{m}\sum_{j=1}^{m} x_0^j + \frac{1}{m}\sum_{j=1}^{m} e_k^j + \frac{1}{m}\sum_{\ell=0}^{k-1}\sum_{j=1}^{m} e_\ell^j. \tag{21}$$

Then, from (20) and (21) we have

$$\begin{aligned}\|x_{k+1}^i - \hat{x}_{k+1}\| \leq{}& \left\|\sum_{j=1}^{m}\left([\Phi(k,0)]_{ij} - \frac{1}{m}\right) x_0^j\right\| + \left\|e_k^i - \frac{1}{m}\sum_{j=1}^{m} e_k^j\right\| \\ &+ \left\|\sum_{\ell=0}^{k-1}\sum_{j=1}^{m}\left([\Phi(k,\ell+1)]_{ij} - \frac{1}{m}\right) e_\ell^j\right\|.\end{aligned}$$

Therefore, for all $j \in V$ and all k,

$$\begin{aligned}\|x_{k+1}^i - \hat{x}_{k+1}\| \leq{}& \sum_{j=1}^{m}\left|[\Phi(k,0)]_{ij} - \frac{1}{m}\right| \|x_0^j\| + \frac{1}{m}\sum_{j\neq i}\|e_k^i - e_k^j\| \\ &+ \sum_{\ell=0}^{k-1}\sum_{j=1}^{m}\left|[\Phi(k,\ell+1)]_{ij} - \frac{1}{m}\right| \|e_\ell^j\|.\end{aligned}$$

At this point, we use the rate of convergence result from Lemma 2 to bound $\left|[\Phi(k,\ell)]_{ij} - \frac{1}{m}\right|$. By doing so, we obtain

$$\|x_{k+1}^i - \hat{x}_{k+1}\| \leq \theta\beta^k \sum_{j=1}^{m}\|x_0^j\| + \frac{1}{m}\sum_{j\neq i}\|e_k^i - e_k^j\| + \theta\sum_{\ell=0}^{k-1}\sum_{j=1}^{m}\beta^{k-(\ell+1)}\|e_\ell^j\|,$$

where $\beta < 1$ (see Lemma 2). By taking the expectation and using (18) we find that for all i and k,

$$\mathbb{E}[\|x_{k+1}^i - \hat{x}_{k+1}\|] \leq \theta \beta^k \sum_{j=1}^m \mathbb{E}[\|x_0^j\|] + \frac{2(m-1)c}{m}\alpha_k + m\theta c \sum_{\ell=0}^{k-1} \beta^{k-(\ell+1)}\alpha_\ell,$$

with $c = (\max_{i \in V} r_i)(C + \nu)/\sigma_x$. Next, we multiply the preceding relation with α_{k+1} and after using $ab \leq (a^2 + b^2)/2$ for the terms $\alpha_{k+1}\alpha_k$ and $\alpha_{k+1}\alpha_\ell$, we obtain

$$\alpha_{k+1}\mathbb{E}[\|x_{k+1}^i - \hat{x}_{k+1}\|] \leq \theta \alpha_{k+1}\beta^k \sum_{j=1}^m \mathbb{E}[\|x_0^j\|] + \frac{(m-1)c}{m}(\alpha_k^2 + \alpha_{k+1}^2)$$
$$+ m\frac{\theta c}{2}\sum_{\ell=0}^{k-1} \beta^{k-(\ell+1)}(\alpha_{k+1}^2 + \alpha_\ell^2),$$

We observe that by $\sum_{k=0}^\infty \alpha_k^2 < \infty$ it follows that $\alpha_k \to 0$, and hence α_k is bounded. This and the fact $\beta < 1$ imply $\sum_{k=0}^\infty \alpha_{k+1}\beta^k < \infty$. The sum $\sum_{k=0}^\infty (\alpha_k^2 + \alpha_{k+1}^2)$ is obviously summable when $\sum_{k=0}^\infty \alpha_k^2 < \infty$. For the last term, by $\beta < 1$ we have

$$\sum_{k=1}^\infty \sum_{\ell=0}^{k-1} \beta^{k-(\ell+1)}(\alpha_{k+1}^2 + \alpha_\ell^2) \leq \sum_{k=1}^\infty \frac{\alpha_{k+1}^2}{1-\beta} + \sum_{k=1}^\infty \sum_{\ell=0}^{k-1} \beta^{k-(\ell+1)}\alpha_\ell^2.$$

The first sum is finite since $\sum_{k=0}^\infty \alpha_k^2 < \infty$, while the second sum is finite by Lemma 3. Thus, $\sum_{k=0}^\infty \alpha_{k+1}\mathbb{E}[\|x_{k+1}^i - \hat{x}_{k+1}\|] < \infty$ for all $i \in V$ and by the monotone convergence theorem [7], it follows $\mathbb{E}\left[\sum_{k=0}^\infty \alpha_{k+1}\|x_{k+1}^i - \hat{x}_{k+1}\|\right] < \infty$. When the expected value of a positive random variable is finite, then the variable must be finite with probability 1, so we must have $\sum_{k=0}^\infty \alpha_{k+1}\|x_{k+1}^i - \hat{x}_{k+1}\| < \infty$ for all $i \in V$, with probability 1.

A similar analysis proves $\sum_{k=0}^\infty \alpha_k|\hat{\eta}_k - \eta_k^i| < \infty$ for all $i \in V$, with probability 1. □

Lemma 4 provides one important property of the iterates generated by our algorithm. We next provide another important relation that in a way captures a descent property of the iterates in terms of the Lyapunov function given by the sum of the Bregman-distance functions B_x and B_η. For notational convenience, we define

$$z_k^i = \begin{bmatrix} x_k^i \\ \eta_k^i \end{bmatrix}, \quad \hat{z}_k = \begin{bmatrix} \hat{x}_k \\ \hat{\eta}_k \end{bmatrix} \qquad \text{for all } k \geq 0. \tag{22}$$

Also, we introduce the notation $\mathbb{E}_k[\cdot] = \mathbb{E}[\cdot \mid F_k]$.

We have the following relation for algorithm (10)–(11), under the assumption that the Bregman-distance functions B_x and B_η are convex in their second arguments.

Lemma 5. *Let Assumptions 1, 2 and 3 hold. Then, for algorithm* (10)–(11) *we have with probability 1 for any* $x^* \in X^*$ *and all* $k \geq 0$,

$$\sum_{i=1}^{m} \mathbb{E}_k \left[B_x(x^*, x_{k+1}^i) + B_\eta(\eta^*, \eta_{k+1}^i) \right] \le \sum_{i=1}^{m} \left(B_x(x^*, x_k^i) + B_\eta(\eta^*, \eta_k^i) \right)$$
$$- \alpha_k \left(\sum_{i=1}^{m} r_i g_i(\hat{z}_k) + \hat{\eta}_k - \eta^* \right) + \alpha_k \bar{r} \left(C \sum_{j=1}^{m} \|\hat{x}_k - x_k^j\| + \sum_{j=1}^{m} |\hat{\eta}_k - \eta_k^j| \right)$$
$$+ \alpha_k^2 m \left(\frac{\bar{r}^2 (C+\nu)^2}{2\sigma_x} + \frac{\left(\frac{1}{m} + \bar{r}\right)^2}{2\sigma_\eta} \right),$$

where $\bar{r} = \max_{i \in V} r_i$.

Proof. By the definition of the Bregman function B_x, we have

$$B_x(x^*, x_{k+1}^i) - B_x(x^*, \tilde{x}_k^i) = w_x(\tilde{x}_k^i) - w_x(x_{k+1}^i) - \langle \nabla w_x(x_{k+1}^i), x^* - x_{k+1}^i \rangle$$
$$+ \langle \nabla w_x(\tilde{x}_k^i), x^* - \tilde{x}_k^i \rangle.$$

Noting that $w_x(\tilde{x}_k^i) - w_x(x_{k+1}^i) = \langle \nabla w_x(\tilde{x}_k^i), \tilde{x}_k^i - x_{k+1}^i \rangle - B_x(x_{k+1}^i, \tilde{x}_k^i)$, we obtain

$$B_x(x^*, x_{k+1}^i) - B_x(x^*, \tilde{x}_k^i) = \langle \nabla w_x(\tilde{x}_k^i) - \nabla w_x(x_{k+1}^i), x^* - x_{k+1}^i \rangle - B_x(x_{k+1}^i, \tilde{x}_k^i). \tag{23}$$

Now, using the optimality condition for (11), since $x^* \in X$, we have $\langle \alpha_k r_i \tilde{d}_k^i + \nabla \omega_x(x_{k+1}^i) - \nabla \omega_x(\tilde{x}_k^i), x^* - x_{k+1}^i \rangle \ge 0$,where $\tilde{d}_k^i = \nabla_x g_i(\tilde{z}_k^i) + \varepsilon_k^i$. This implies

$$\langle \nabla \omega_x(\tilde{x}_k^i) - \nabla \omega_x(x_{k+1}^i), x^* - x_{k+1}^i \rangle \le \alpha_k r_i \langle \tilde{d}_k^i, x^* - x_{k+1}^i \rangle.$$

Upon substituting the preceding inequality in (23), we obtain

$$B_x(x^*, x_{k+1}^i) - B_x(x^*, \tilde{x}_k^i) \le \alpha_k r_i \langle \tilde{d}_k^i, x^* - x_{k+1}^i \rangle - B_x(x_{k+1}^i, \tilde{x}_k^i)$$
$$\le \alpha_k r_i \langle \tilde{d}_k^i, x^* - \tilde{x}_k^i \rangle + \alpha_k r_i \langle \tilde{d}_k^i, \tilde{x}_k^i - x_{k+1}^i \rangle - \frac{\sigma_x}{2} \|x_{k+1}^i - \tilde{x}_k^i\|^2. \tag{24}$$

where the last inequality follows from the strong convexity of the Bregman function, i.e., $B_x(x_{k+1}^i, \tilde{x}_k^i) \ge \frac{\sigma_x}{2} \|x_{k+1}^i - \tilde{x}_k^i\|^2$. Further, by Hölder's inequality we have $\alpha_k r_i \langle \tilde{d}_k^i, \tilde{x}_k^i - x_{k+1}^i \rangle \le \alpha_k r_i \|\tilde{d}_k^i\|_* \|\tilde{x}_k^i - x_{k+1}^i\|$. Using this and the scalar inequality $2ab \le a^2 + b^2$, we obtain

$$\alpha_k r_i \langle \tilde{d}_k^i, \tilde{x}_k^i - x_{k+1}^i \rangle \le 2 \frac{\alpha_k r_i}{\sqrt{2\sigma_x}} \|\tilde{d}_k^i\|_* \frac{\sqrt{\sigma_x}}{\sqrt{2}} \|\tilde{x}_k^i - x_{k+1}^i\|$$
$$\le \alpha_k^2 r_i^2 \frac{\|\tilde{d}_k^i\|_*^2}{2\sigma_x} + \frac{\sigma_x}{2} \|x_{k+1}^i - \tilde{x}_k^i\|^2. \tag{25}$$

Upon combining (25) and (24) we arrive at

$$B_x(x^*, x_{k+1}^i) - B_x(x^*, \tilde{x}_k^i) \le \alpha_k r_i \langle \tilde{d}_k^i, x^* - \tilde{x}_k^i \rangle + \alpha_k^2 r_i^2 \frac{\|\tilde{d}_k^i\|_*^2}{2\sigma_x}.$$

Since $\tilde{d}_k^i = \nabla_x g_i(\tilde{z}_k^i) + \varepsilon_k^i$, by taking the conditional expectation with respect to the history F_k, and by using (13) and $\mathbb{E}_k[\tilde{d}_k^i] = \nabla_x g_i(v_k^i)$, we obtain

$$\mathbb{E}_k\left[B_x(x^*, x_{k+1}^i)\right] \le B_x(x^*, \tilde{x}_k^i) - \alpha_k r_i \langle \nabla_x g_i(\tilde{z}_k^i), \tilde{x}_k^i - x^* \rangle + \alpha_k^2 r_i^2 \frac{(C+\nu)^2}{2\sigma_x}. \tag{26}$$

Proceeding similarly, we can derive the following inequality for the iterates involving the min-max value estimates η_k^i,

$$\mathbb{E}_k\left[B_\eta(\eta^*, \eta_{k+1}^i)\right] \le B_\eta(\eta^*, \tilde{\eta}_k^i) - \alpha_k \left(\frac{1}{m} + r_i \nabla_\eta g_i(\tilde{z}_k^i)\right)(\tilde{\eta}_k^i - \eta^*) + \alpha_k^2 \frac{\left(\frac{1}{m} + r_i\right)^2}{2\sigma_\eta}, \tag{27}$$

where we use the fact $|\nabla_\eta g_i(\tilde{z}_k^i)| \le 1$. By the convexity of g_i and the subgradient property, we have

$$\nabla g_i(\tilde{z}_k^i)'(\tilde{z}_k^i - z^*) \ge g_i(\tilde{z}_k^i) - g_i(z^*) = g_i(\tilde{z}_k^i).$$

where the equality follows by $g_i(z^*) = \max\{0, f_i(x^*) - \eta^*\}$ and $f_i(x^*) - \eta^* \le 0$ for all $i \in V$ and any optimal point $x^* \in X^*$. Upon adding the inequalities (26) and (27), and using the convexity of g_i, we obtain for any $x^* \in X^*$ and all $k \ge 0$ and $i \in V$,

$$\begin{aligned}\mathbb{E}_k\left[B_x(x^*, x_{k+1}^i) + B_\eta(\eta^*, \eta_{k+1}^i)\right] &\le B_x(x^*, \tilde{x}_k^i) + B_\eta(\eta^*, \tilde{\eta}_k^i) - \alpha_k \frac{1}{m}(\tilde{\eta}_k^i - \eta^*) \\ &- \alpha_k r_i g_i(\tilde{z}_k^i) + \alpha_k^2 \left(\frac{r_i^2 (C+\nu)^2}{2\sigma_x} + \frac{\left(\frac{1}{m} + r_i\right)^2}{2\sigma_\eta}\right).\end{aligned} \tag{28}$$

Now, by Assumption 3 on the convexity of the Bregman functions B_x and B_η and the doubly stochasticity of the weight matrix W_k (Assumption 1-b), we have

$$\sum_{i=1}^m B_x(x^*, \tilde{x}_k^i) = \sum_{i=1}^m B_x\left(x^*, \sum_{j=1}^m [W_k]_{ij} x_k^j\right) \le \sum_{i=1}^m \sum_{j=1}^m [W_k]_{ij} B_x(x^*, x_k^j) = \sum_{j=1}^m B_x(x^*, x_k^j).$$

Similarly, we have $\sum_{i=1}^m B_\eta(\eta^*, \tilde{\eta}_k^i) \le \sum_{j=1}^m B_\eta(\eta^*, \eta_k^j)$. Thus, by summing inequalities in (28) over $i \in V$ we obtain for any $x^* \in X^*$ and all $k \ge 0$,

$$\begin{aligned}\sum_{i=1}^m \mathbb{E}_k\left[B_x(x^*, x_{k+1}^i) + B_\eta(\eta^*, \eta_{k+1}^i)\right] &\le \sum_{i=1}^m \left(B_x(x^*, x_k^i) + B_\eta(\eta^*, \eta_k^i)\right) \\ - \alpha_k \frac{1}{m} \sum_{i=1}^m (\tilde{\eta}_k^i - \eta^*) - \alpha_k \sum_{i=1}^m r_i g_i(\tilde{z}_k^i) &+ \alpha_k^2 m \left(\frac{\bar{r}^2 (C+\nu)^2}{2\sigma_x} + \frac{\left(\frac{1}{m} + \bar{r}\right)^2}{2\sigma_\eta}\right),\end{aligned} \tag{29}$$

where $\bar{r} = \max_{i \in V} r_i$.

From the definition of $\tilde{\eta}_k^i$ and the doubly stochasticity of the matrix W_k, we have $\sum_{i=1}^m \tilde{\eta}_k^i = \sum_{i=1}^m \sum_{j=1}^m [W_k]_{ij}\, \eta_k^j = \sum_{j=1}^m \eta_k^j = m\hat{\eta}_k$. Using this identity, together with adding and subtracting the term $\alpha_k \sum_{i=1}^m r_i g_i(\hat{z}_k)$, from the preceding relation we obtain

$$\sum_{i=1}^m \mathbb{E}_k \left[B_x(x^*, x_{k+1}^i) + B_\eta(\eta^*, \eta_{k+1}^i)\right] \leq \sum_{i=1}^m \left(B_x(x^*, x_k^i) + B_\eta(\eta^*, \eta_k^i)\right)$$
$$- \alpha_k \left(\sum_{i=1}^m r_i g_i(\hat{z}_k) + \hat{\eta}_k - \eta^*\right) + \alpha_k \sum_{i=1}^m r_i \left|g_i(\hat{z}_k) - g_i(\tilde{z}_k^i)\right|$$
$$+ \alpha_k^2 m \left(\frac{\bar{r}^2 (C+\nu)^2}{2\sigma_x} + \frac{\left(\frac{1}{m} + \bar{r}\right)^2}{2\sigma_\eta}\right), \tag{30}$$

where $\hat{z}_k$ is as defined in (22). Next, we consider the term $\sum_{i=1}^m r_i \left|g_i(\hat{z}_k) - g_i(\tilde{z}_k^i)\right|$. By the definition of g_i and relation $|\max\{a,0\} - \max\{b,0\}| \leq |a-b|$ valid for any two scalars a and b, we have the following:

$$\left|g_i(\hat{z}_k) - g_i(\tilde{z}_k^i)\right| \leq \left|f_i(\hat{x}_k) - f_i(\tilde{x}_k^i) - \hat{\eta}_k + \tilde{\eta}_k^i\right| \leq C\|\hat{x}_k - \tilde{x}_k^i\| + \left|\hat{\eta}_k - \tilde{\eta}_k^i\right|,$$

where in the first inequality we use the definition of $\tilde{z}_k^i$ in (10), while in the last inequality we use the subgradient boundedness assumption for f_i. Further, by using the definition of the variables $\tilde{x}_k^i$ and $\tilde{\eta}_k^i$ in (10), the stochasticity of W_k and the convexity of the norm, we obtain

$$\left|g_i(\hat{z}_k) - g_i(\tilde{z}_k^i)\right| \leq \sum_{j=1}^m [W_k]_{ij} \left[C\|\hat{x}_k - x_k^j\| + |\hat{\eta}_k - \eta_k^j|\right].$$

Therefore, by using the doubly stochasticity of W_k and $\bar{r} = \max r_i$, we obtain

$$\sum_{i=1}^m r_i \left|g_i(\hat{z}_k) - g_i(\tilde{z}_k^i)\right| \leq \bar{r}C \sum_{j=1}^m \|\hat{x}_k - x_k^j\| + \bar{r} \sum_{j=1}^m |\hat{\eta}_k - \eta_k^j|.$$

Substituting this estimate back in (30) we get the desired result. □

We are now ready to prove our main convergence result. The result essentially states that under suitable conditions on the step size α_k, all the agents' estimates converge to a common optimal point. Moreover, the agents' estimates of the min-max value also converge to the optimal value of the problem.

Theorem 3. *Let Assumptions 1, 2, and 3 hold. Let the step sizes satisfy $\sum_{k=0}^\infty \alpha_k = \infty$ and $\sum_{k=0}^\infty \alpha_k^2 < \infty$. Then, for all $i \in V$, the agents' iterates x_k^i and η_k^i generated by algorithm* (10)–(11) *are such that with probability 1 for all $i \in V$:*

(a) The decision variables x_k^i converge to a common optimal (random) point $x^* \in X^*$.
(b) The estimates η_k^i converge to the optimal value η^* of the min-max problem.

Proof. Our analysis is based on applying the Robbins-Siegmund result from Theorem 2 to the inequality derived in Lemma 5. By our assumption on the step sizes we have $\sum_{k=0}^{\infty} \alpha_k^2 < \infty$, which trivially implies that

$$\sum_{k=0}^{\infty} \alpha_k^2 m \left(\frac{\bar{r}^2 (C+\nu)^2}{2\sigma_x} + \frac{\left(\frac{1}{m} + \bar{r}\right)^2}{2\sigma_\eta} \right) < \infty.$$

Furthermore, by virtue of Lemma 4 we have

$$\sum_{k=0}^{\infty} \alpha_k \bar{r} \sum_{j=1}^{m} \left(C \|\hat{x}_k - x_k^j\| + |\hat{\eta}_k - \eta_k^j| \right) < \infty. \tag{31}$$

In addition, by the definition $\hat{z}_k^i$ is a convex combination of $(x_k^i, \eta_k^i) \in X \times \mathbb{R}$ for all $i \in V$ and k, implying by Lemma 1 that $\sum_{i=1}^{m} r_i g_i(\hat{z}_k) + \hat{\eta}_k - \eta^* \geq 0$ for all $k \geq 0$. Thus, we can apply Lemma 2 to the relation of Lemma 5 and infer that with probability 1, $\sum_{i=1}^{m} \left(B_x(x^*, x_k^i) + B_\eta(\eta^*, \eta_k^i) \right)$ converges for every $x^* \in X^*$ and

$$\sum_{k=0}^{\infty} \alpha_k \left(\sum_{i=1}^{m} r_i g_i(\hat{z}_k) + \hat{\eta}_k - \eta^* \right) < \infty. \tag{32}$$

Now, since $\sum_{k=0}^{\infty} \alpha_k = \infty$, from (31)–(32) it follows that there exists a subsequence indexed by $\{k_\ell\}$ such that with probability 1,

$$\lim_{\ell \to \infty} \left(\sum_{i=1}^{m} r_i g_i(\hat{z}_{k_\ell}) + \hat{\eta}_{k_\ell} - \eta^* \right) = 0,$$

$$\lim_{\ell \to \infty} \|\hat{x}_{k_\ell} - x_{k_\ell}^j\| = 0 \text{ and } \lim_{\ell \to \infty} |\hat{\eta}_{k_\ell} - \eta_{k_\ell}^j| = 0 \text{ for all } j. \tag{33}$$

Since $\sum_{i=1}^{m} \left(B_x(x^*, x_k^i) + B_\eta(\eta^*, \eta_k^i) \right)$ converges for every $x^* \in X^*$, the sequence $\sum_{i=1}^{m} \left(B_x(x^*, x_k^i) + B_\eta(\eta^*, \eta_k^i) \right)$ must be bounded for every $x^* \in X^*$ with probability 1. Note that from Assumption 3 on the convexity of Bregman functions B_x and B_η and their inherent strong convexity property we have

$$\begin{aligned} \frac{1}{m} \sum_{i=1}^{m} \left(B_x(x^*, x_k^i) + B_\eta(\eta^*, \eta_k^i) \right) &\geq B_x(x^*, \hat{x}_k) + B_\eta(\eta^*, \hat{\eta}_k) \\ &\geq \frac{\sigma_x}{2} \|x^* - \hat{x}_k\|^2 + \frac{\sigma_\eta}{2} |\eta^* - \hat{\eta}_k|^2. \end{aligned}$$

Thus, the sequences $\{\hat{x}_k\}$ and $\{\hat{\eta}_k\}$ are also bounded with probability 1. Hence, along a further subsequence, which without loss of generality we can let it be indexed by the same index set $\{k_\ell, \ell = 1, 2, \ldots\}$, with probability 1 we have $\lim_{\ell \to \infty} \hat{x}_{k_\ell} = \bar{x}$ and $\lim_{\ell \to \infty} \hat{\eta}_{k\ell} = \bar{\eta}$, where $\bar{x} \in X$ since X is closed. Moreover, with probability 1 the limit points satisfy

$$\sum_{i=1}^{m} r_i g_i(\bar{x}) + \bar{\eta} - \eta^* = 0.$$

From this relation and Lemma 1 it follows that $\bar{x} \in X^*$ and $\bar{\eta} = \eta^*$ with probability 1.

In view of (33), and $\hat{x}_{k_\ell} \to \bar{x}$ and $\hat{\eta}_{k_\ell} \to \eta^*$, we further have $x^j_{k_\ell} \to \bar{x}$ and $\eta^j_{k_\ell} \to \bar{\eta}$ for all j with probability 1. Therefore, $\lim_{\ell\to\infty} \sum_{i=1}^m \left(B_x(\bar{x}, x^i_{k_\ell}) + B_\eta(\eta^*, \eta^i_{k_\ell})\right) = 0$. However, we have established that the sequence $\sum_{i=1}^m (B_x(x^*, x^i_k) + B_\eta(\eta^*, \eta^i_k))$ converges with probability 1 for any $x^* \in X^*$, which in view of $\bar{x} \in X^*$, implies that with probability 1,

$$\lim_{k\to\infty} \sum_{i=1}^m (B_x(\bar{x}, x^i_k) + B_\eta(\eta^*, \eta^i_k)) = 0.$$

Finally, by the strong convexity of the Bregman-distance functions B_x and B_η, it follows that $x^j_k \to \bar{x}$ and $\eta^j_k \to \eta^*$ with probability 1 for all j. □

One may extend algorithm (10)–(11) to the case when each agent i uses a different Bregman function $B_{\eta,i}$ instead of a common one B_η. Following the same analysis, with a slight modification, it can be seen that the convergence result of Theorem 3 would be applicable to such a modified algorithm.

4 Primal-Dual Approach

In this section we present a distributed primal-dual algorithm which is motivated by the classical work of Arrow-Hurwicz-Uzawa [2]. Recently, a primal-dual method was studied in [24] for approximate solutions to saddle-point problems by considering standard Euclidean norm. The use of Bregman distance for a saddle-point problem was studied in [26]. The prior work is dealing with centralized problems, while here we consider a primal-dual method with Bregman distances for solving the *distributed min-max problem* in its epigraph formulation (3).

In order to apply primal-dual approach, we need to slightly modify the original epigraph formulation (3) of the min-max problem but without changing its set of optimal solutions. This is needed to ensure the convergence of the primal-dual algorithm. Specifically, we assume that we have a closed convex interval D such that $\eta^* \in D$, which is equivalent to having some (arbitrarily large) upper and lower estimates for η^*. Having such a set, we consider a modification of epigraph formulation (3), given by

$$\begin{array}{lll} \text{minimize} & \eta & \\ \text{subject to} & f_i(x) \le \eta & \text{for all } x \in X,\ \eta \in D, \text{ and } i \in V. \end{array} \tag{34}$$

This problem has the same solutions as the original min-max problem since $\eta^* \in D$. The Lagrangian function associated with this problem is given by

$$\mathcal{L}(x, \eta, \mu) = \sum_{i=1}^m \mu_i (f_i(x) - \eta) + \eta \quad \text{for } x \in X, \eta \in D, \mu \ge 0.$$

We can restrict the domain of dual variables of the Lagrangian function to the unit interval, as we know that the dual optimal multipliers satisfy $\sum_{i=1}^{m} \mu_i^* = 1$, as seen in Section 3. Thus, we will consider the Lagrangian function with a restricted domain (yet large enough to contain all dual optimal solutions):

$$\mathscr{L}(x,\eta,\mu) = \sum_{i=1}^{m} \mu_i (f_i(x) - \eta) + \eta \quad \text{for } x \in X, \eta \in D, \mu \in I^m, \tag{35}$$

where I is the interval $[0,1]$ and I^m denotes the product of m copies of I. We are interested in determining a primal-dual optimal pair for problem (34) by an algorithm aimed at finding a saddle-point of the reduced Lagrangian (35).

4.1 Primal-Dual Algorithm

We are interested in distributed algorithm for computing a saddle-point of the Lagrangian (35). To accommodate distributed computations among m agents, we write the Lagrangian function as a sum of m functions, as follows:

$$L(x,\eta,\mu) = \sum_{i=1}^{m} \mu_i (f_i(x) - \eta) + \eta = \sum_{i=1}^{m} \left(\mu_i (f_i(x) - \eta) + \frac{1}{m}\eta \right). \tag{36}$$

Lagrangian-function component $L_i(x,\eta,\mu_i) = \mu_i (f_i(x) - \eta) + \frac{1}{m}\eta$ is assigned to agent $i \in V$ for processing without sharing the information about the function with any other agent.

The distributed algorithm will use the Bregman-distance functions B_x and B_η, as introduced in Section 3.2. In addition, we introduce another collection of Bregman-distance functions, one for each of the agents in order to handle the Lagrange multipliers for its constraint set $\{x \mid f_i(x) \leq \eta\}$. For this, for each $i \in V$, we let $B_{\mu,i}(\cdot,\cdot)$ be a Bregman-distance function associated with a strongly convex function $\omega_{\mu,i} : \mathbb{R} \to \mathbb{R}$ with parameter $\sigma_{\mu,i} > 0$.

The proposed primal-dual distributed algorithm for finding a saddle-point of the Lagrangian is as follows. At every iteration k, each agent i has estimates x_k^i, η_k^i and μ_i^k, respectively, for an optimal solution of the min-max problem, the optimal value and the Lagrangian multiplier associated with the constraint $\{x \mid f_i(x) \leq \eta\}$ that agent i is responsible for. Every agent firstly performs an intermittent adjustment of the variables x_k^i and η_k^i as in (10) to obtain the estimates $\tilde{x}_k^i$ and $\tilde{\eta}_i^k$. For convenience, we restate these updates:

$$\begin{bmatrix} \tilde{x}_k^i \\ \tilde{\eta}_k^i \end{bmatrix} = \sum_{j=1}^{m} [W_k]_{ij} \begin{bmatrix} x_k^j \\ \eta_k^j \end{bmatrix}. \tag{37}$$

This step ensures that agents locally align the variables that are coupling. We note that the agents do not have an intermittent adjustment for their multiplier estimates μ_k^i, as these variables do not couple.

Then, every agent $i \in V$ generates new iterates by taking step toward minimizing its own Lagrangian function $L_i(x,\eta,\mu_i) = \mu_i\,(f_i(x) - \eta) + \frac{1}{m}\eta$ with respect to (x,η) and maximizing it with respect to μ_i, in the following manner:

$$\begin{aligned} x_{k+1}^i &= \operatorname{argmin}_{y\in X}\left[\alpha_k \mu_k^i \langle d_k^i + \varepsilon_k^i, y\rangle + B_x(y, \tilde{x}_k^i)\right], \\ \eta_{k+1}^i &= \operatorname{argmin}_{s\in D}\left[\alpha_k (1/m - \mu_k^i) s + B_\eta(s, \tilde{\eta}_k^i)\right], \\ \mu_{k+1}^i &= \operatorname{argmin}_{\zeta\in I}\left[\alpha_k \left(\tilde{\eta}_k^i - f_i(\tilde{x}_k^i)\right)\zeta + B_{\mu,i}(\zeta, \mu_k^i)\right], \end{aligned} \tag{38}$$

where d_k^i is a subgradient of $f_i(x)$ evaluated at $\tilde{x}_k^i$ and ε_k is a random error in the subgradient evaluation. For each i, the initial values $x_0^i \in X$, $\eta_0^i \in D$, and $\mu_0^i \in I$ are random and independent from the stochastic errors ε_k.

4.2 Algorithm Convergence

The analysis is similar to that of algorithm (10)–(11) using exact-penalty approach. First, we state a lemma which relates the local iterates of the primal-dual algorithm to their respective average trajectory for the coupling variables x_k^i and η_k^i.

Lemma 6. *Let Assumptions 1 and 2 hold, and let the step sizes satisfy $\sum_{k=0}^{\infty}\alpha_k^2 < \infty$. Then, for the instantaneous averages $\hat{x}_k$ and $\hat{\eta}_k$ of the iterates generated by algorithm* (37)–(38)*, we have $\sum_{k=0}^{\infty}\alpha_k\|\hat{x}_k - x_k^i\| < \infty$ and $\sum_{k=0}^{\infty}\alpha_k|\hat{\eta}_k - \eta_k^i| < \infty$ for all $i \in V$, with probability 1.*

Proof. The proof is similar to that of Lemma 4 and uses relations $\mu_k^i \in I$, $i \in V$.

As a Lyapunov function, we choose the composite function

$$\mathbf{B}(z,\mu,\mathbf{z}_k,\mu_k) = \sum_{i=1}^{m}\left[B_x(x,x_k^i) + B_\eta(\eta,\eta_k^i) + B_{\mu,i}(\mu_i,\mu_k^i)\right], \tag{39}$$

where $\mathbf{z}_k = (x_k^1,\ldots,x_k^m,\eta_k^1,\ldots,\eta_k^m)$ and $\mu_k = (\mu_k^1,\ldots,\mu_k^m)$. We next establish a descent-type relation for the expected value of the Lyapunov function, which requires an appropriate *sigma*-field. We define the σ-field as follows:

$$F_k = \{x_t^i, \eta_t^i, \mu_t^i, i \in V, t = 0,1,\ldots,k\} \qquad \text{for } k \geq 0.$$

From now on, we abbreviate the conditional expectation notation by $\mathbb{E}_k[\cdot] = \mathbb{E}[\cdot \mid F_k]$. We have the following result that will play the key role in the convergence analysis.

Lemma 7. *Let $D \subset \mathbb{R}$ be a convex compact set such that $\eta^* \in D$. Let Assumptions 1, 2 and 3 hold. Then, for algorithm* (37)–(38) *the following relation holds with probability 1, for all $k \geq 0$, all $z^* = (x^*,\eta^*)$ with $x^* \in X^*$, and all optimal multipliers μ^*,*

$$\mathbb{E}_k[\boldsymbol{B}(z^*,\mu^*,z_{k+1},\mu_{k+1})] \leq \boldsymbol{B}(z^*,\mu^*,z_k,\mu_k) + 2\alpha_k \sum_{j=1}^{m} \left(C\|x_k^j - \hat{x}_k\| + |\eta_k^j - \hat{\eta}_k| \right)$$
$$- \alpha_k \left(\mathscr{L}(\hat{x}_k, \hat{\eta}_k, \mu^*) - \mathscr{L}(x^*, \eta^*, \mu_k) \right)$$
$$+ \alpha_k^2 \left(m\frac{(C+\nu)^2}{2\sigma_x} + \sum_{i=1}^{m} \frac{(1/m - \mu_k^i)^2}{2\sigma_\eta} + \sum_{i=1}^{m} \frac{(f_i(\tilde{x}_k^i) - \tilde{\eta}_k^i)^2}{2\sigma_{\mu,i}} \right).$$

Proof. Let $z = (x^*, \eta^*) \in X^* \times D$ and $\mu^* \in I^m$ be arbitrary optimal primal-dual pairs, where I^m is the product of m copies of the interval $I = [0,1]$. Using the optimality conditions for (38) and proceeding as in the proof of Lemma 5, we obtain with probability 1,

$$\mathbb{E}_k\left[B_x(x^*, x_{k+1}^i)\right] \leq B_x(x^*, \tilde{x}_k^i) - \alpha_k \mu_k^i \langle \mathbb{E}_k[d_k^i + \varepsilon_k^i], \tilde{x}_k^i - x^* \rangle + \alpha_k^2 \frac{(C+\nu)^2}{2\sigma_x}.$$

Now, since $\mathbb{E}_k[\varepsilon_k^i] = 0$, and d_k^i is a subgradient of $f_i(x)$ at $\tilde{x}_k^i$ it follows that

$$\mathbb{E}_k\left[B_x(x^*, x_{k+1}^i)\right] \leq B_x(x^*, \tilde{x}_k^i) - \alpha_k \mu_k^i (f_i(\tilde{x}_k^i) - f_i(x^*)) + \alpha_k^2 \frac{(C+\nu)^2}{2\sigma_x}. \tag{40}$$

A similar analysis gives the following inequality for the iterates η_k^i,

$$B_\eta(\eta^*, \eta_{k+1}^i) \leq B_\eta(\eta^*, \tilde{\eta}_k^i) - \alpha_k (1/m - \mu_k^i)(\tilde{\eta}_k^i - \eta^*) + \alpha_k^2 \frac{(1/m - \mu_k^i)^2}{2\sigma_\eta}, \tag{41}$$

and the inequality for the multiplier iterates μ_k^i,

$$B_{\mu,i}(\mu_i^*, \mu_{k+1}^i) \leq B_{\mu,i}(\mu_i^*, \mu_k^i) + \alpha_k (f_i(\tilde{x}_k^i) - \tilde{\eta}_k^i)(\mu_k^i - \mu_i^*) + \alpha_k^2 \frac{(f_i(\tilde{x}_k^i) - \tilde{\eta}_k^i)^2}{2\sigma_{\mu,i}}. \tag{42}$$

Summing equations (40)–(42), and then summing the resulting relation over all $i \in V$, we have with probability 1,

$$\mathbb{E}_k[\mathbf{B}(z^*, \mu^*, \mathbf{z}_{k+1}, \mu_{k+1})] \leq \mathbf{B}(z^*, \mu^*, \tilde{\mathbf{z}}_k, \mu_k) + \alpha_k^2 K \tag{43}$$
$$- \alpha_k \sum_{i=1}^{m} \left(\underbrace{\mu_k^i (f_i(\tilde{x}_k^i) - f_i(x^*))}_{\text{Term 1}} + \underbrace{(1/m - \mu_k^i)(\tilde{\eta}_k^i - \eta^*)}_{\text{Term 2}} - \underbrace{(f_i(\tilde{x}_k^i) - \tilde{\eta}_k^i)(\mu_k^i - \mu_i^*)}_{\text{Term 3}} \right),$$

where we use notation (39) for the Lyapunov function, $\tilde{\mathbf{z}}_k = (\tilde{x}_k^1, \ldots, \tilde{x}_k^m, \tilde{\eta}_k^1, \ldots, \tilde{\eta}_k^m)$, and $K = m\frac{(C+\nu)^2}{2\sigma_x} + \sum_{i=1}^m \frac{(1/m-\mu_k^i)^2}{2\sigma_\eta} + \sum_{i=1}^m \frac{(f_i(\tilde{x}_k^i)-\tilde{\eta}_k^i)^2}{2\sigma_{\mu,i}}$. We now estimate the identified terms in the preceding relation by adding and subtracting $f(\hat{x}_k)$ or $\hat{\eta}_k$. We have

$$f_i(x_k^i) - f_i(x^*) = (f_i(\hat{x}_k) - f_i(x^*)) + (f_i(\tilde{x}_k^i) - f_i(\hat{x}_k)) \geq f_i(\hat{x}_k) - f_i(x^*) - C\|\tilde{x}_k^i - \hat{x}_k\|,$$

where the inequality follows from the convexity of f_i and subgradient boundedness. Since $\mu_k^i \in I = [0,1]$ for all $i \in V$, it follows

$$\text{Term } 1 \geq \mu_k^i(f_i(\hat{x}_k) - f_i(x^*)) - C\|\tilde{x}_k^i - \hat{x}_k\|. \tag{44}$$

For the second term, we have

$$\begin{aligned}(1/m - \mu_k^i)(\tilde{\eta}_k^i - \eta^*) &= (1/m - \mu_k^i)(\hat{\eta}_k - \eta^*) + (1/m - \mu_k^i)(\tilde{\eta}_k^i - \hat{\eta}_k) \\ &\geq (1/m - \mu_k^i)(\hat{\eta}_k - \eta^*) - |1/m - \mu_k^i|\,|\tilde{\eta}_k^i - \hat{\eta}_k|.\end{aligned}$$

Using $\mu_k^i \in I = [0,1]$ for all $i \in V$, we obtain

$$\text{Term } 2 \geq (1/m - \mu_k^i)(\hat{\eta}_k - \eta^*) - |\tilde{\eta}_k^i - \hat{\eta}_k|. \tag{45}$$

For the third term, we write

$$\begin{aligned}(f_i(\tilde{x}_k^i) - \tilde{\eta}_k^i)(\mu_k^i - \mu_i^*) &= (f_i(\hat{x}_k) - \hat{\eta}_k)(\mu_k^i - \mu_i^*) \\ &\quad + \left((f_i(\tilde{x}_k^i) - f_i(\hat{x}_k)) - (\tilde{\eta}_k^i - \hat{\eta}_k)\right)(\mu_k^i - \mu_i^*) \\ &\geq (f_i(\hat{x}_k) - \hat{\eta}_k)(\mu_k^i - \mu_i^*) - \left(C\|\tilde{x}_k^i - \hat{x}_k\| + |\tilde{\eta}_k - \hat{\eta}_k|\right)|\mu_k^i - \mu_i^*|,\end{aligned}$$

where the inequality follows by the convexity of f_i and the subgradient boundedness. Again, since $\mu_k^i, \mu \in I$, we see that $|\mu_k^i - \mu_i| \leq 1$, implying

$$\text{Term } 3 \geq (f_i(\hat{x}_k) - \hat{\eta}_k)(\mu_k^i - \mu_i^*) - \left(C\|\tilde{x}_k^i - \hat{x}_k\| + |\tilde{\eta}_k - \hat{\eta}_k|\right). \tag{46}$$

Substituting estimates (44)–(46) back in relation (43), we obtain

$$\begin{aligned}\mathbb{E}_k[\mathbf{B}(z^*, \mu^*, \mathbf{z}_{k+1}, \mu_{k+1})] &\leq \mathbf{B}(z^*, \mu^*, \tilde{\mathbf{z}}_k, \mu_k) + \alpha_k^2 K + 2\alpha_k \sum_{i=1}^m \left(C\|\tilde{x}_k^i - \hat{x}_k\| + |\tilde{\eta}_k^i - \hat{\eta}_k|\right) \\ &- \alpha_k \sum_{i=1}^m \left(\mu_k^i(f_i(\hat{x}_k) - f_i(x^*)) + (1/m - \mu_k^i)(\hat{\eta}_k - \eta^*) - (f_i(\hat{x}_k) - \hat{\eta}_k)(\mu_k^i - \mu_i^*)\right).\end{aligned} \tag{47}$$

By definition in (10), the estimates $\tilde{x}_k^i$ and $\tilde{\eta}_k^i$ are convex combinations of x_k^j and η_k^j, respectively. Since under our assumption the Bregman-distance functions B_x and B_η are convex in the second argument, it follows

$$B_x(x^*, \tilde{x}_k^i) \leq \sum_{j=1}^m [W_k]_{ij} B_x(x^*, x_k^j), \qquad B_\eta(\eta^*, \tilde{\eta}_k^i) \leq \sum_{j=1}^m [W_k]_{ij} B_\eta(\eta^*, \eta_k^j).$$

By summing these relations over $i \in V$ and using the doubly stochasticity of the weight matrix W_k, we obtain

$$\sum_{i=1}^m B_x(x^*, \tilde{x}_k^i) \leq \sum_{j=1}^m B_x(x^*, x_k^j), \qquad \sum_{i=1}^m B_\eta(\eta^*, \tilde{\eta}_k^i) \leq \sum_{j=1}^m B_\eta(\eta^*, \eta_k^j).$$

Using the definition of **B** (see (39)) and these relations, from (47) we have

$$\mathbb{E}_k[\mathbf{B}(z^*,\mu^*,\mathbf{z}_{k+1},\mu_{k+1})] \leq \mathbf{B}(z^*,\mu^*,\mathbf{z}_k,\mu_k) + \alpha_k^2 K + 2\alpha_k \sum_{i=1}^m \left(C\|\tilde{x}_k^i - \hat{x}_k\| + |\tilde{\eta}_k^i - \hat{\eta}_k|\right)$$
$$- \alpha_k \underbrace{\sum_{i=1}^m \left(\mu_k^i(f_i(\hat{x}_k) - f_i(x^*)) + (1/m - \mu_k^i)(\hat{\eta}_k - \eta^*) - (f_i(\hat{x}_k) - \hat{\eta}_k)(\mu_k^i - \mu_i^*)\right)}_{\text{Term}}. \tag{48}$$

Now, we consider the identified term in (48). We note that

$$\sum_{i=1}^m \mu_k^i(f_i(\hat{x}_k) - f_i(x^*)) + (1/m - \mu_k^i)(\hat{\eta}_k - \eta^*) = \mathscr{L}(\hat{x}_k,\hat{\eta}_k,\mu_k) - \mathscr{L}(x^*,\eta^*,\mu_k),$$

$$\sum_{i=1}^m (f_i(\hat{x}_k) - \hat{\eta}_k)(\mu_k^i - \mu_i^*) = \mathscr{L}(\hat{x}_k,\hat{\eta}_k,\mu_k) - \mathscr{L}(\hat{x}_k,\hat{\eta}_k,\mu^*),$$

which imply

$$\text{Term} = \mathscr{L}(\hat{x}_k,\hat{\eta}_k,\mu^*) - \mathscr{L}(\hat{x}^*,\eta^*,\mu_k). \tag{49}$$

Furthermore, from convexity of the norm and the absolute value functions, since W_k is doubly stochastic, and $\tilde{x}_k^i = \sum_{j=1}^m x_k^j$ and $\tilde{\eta}_k^i = \sum_{j=1}^m x_k^j$ it follows that

$$\sum_{i=1}^m \left(C\|\tilde{x}_k^i - \hat{x}_k\| + |\tilde{\eta}_k^i - \hat{\eta}_k|\right) \leq \sum_{j=1}^m \left(C\|x_k^j - \hat{x}_k\| + |\eta_k^j - \hat{\eta}_k|\right). \tag{50}$$

Using (49)–(50) and $K = m\frac{(C+\nu)^2}{2\sigma_x} + \sum_{i=1}^m \frac{(1/m-\mu_k^i)^2}{2\sigma_\eta} + \sum_{i=1}^m \frac{(f_i(\tilde{x}_k^i)-\tilde{\eta}_k^i)^2}{2\sigma_{\mu,i}}$, from (48) we obtain the desired relation. □

In order to connect the limiting vector $(\bar{x},\bar{\eta},\bar{\mu})$ of the iterates generated by the primal-dual algorithm to the solutions of problem (34), we will invoke the necessary and sufficient Karush-Khun-Tucker (KKT) optimality conditions. These conditions are stated below for convenience, as adjusted to our problem.

Theorem 4. *The vector $(\bar{x},\bar{\eta},\bar{\mu})$ is a primal-dual optimal vector if and only if $f_i(\bar{x}) \leq \bar{\eta}$ and $\bar{\mu} \in I^m$ and the following two conditions are satisfied*

$$(\bar{x},\bar{\eta},\bar{\mu}) \in \operatorname{argmin}_{(x,\eta)\in X\times D} \mathscr{L}(x,\eta,\bar{\mu}) \qquad \bar{\mu} \in \operatorname{argmax}_{\mu\in I^m} \mathscr{L}(\bar{x},\bar{\eta},\mu).$$

We are now in position to assert the convergence property of the algorithm for a diminishing step size.

Theorem 5. *Let Assumptions 1, 2, and 3 hold, except for subgradient norm boundedness of Assumption 2-b. Assume that X and D are compact convex sets, and that min-max problem* (3) *has a unique optimal solution x^*. Let the step sizes satisfy $\sum_{k=0}^\infty \alpha_k = \infty$ and $\sum_{k=0}^\infty \alpha_k^2 < \infty$. Then, the agents' iterates x_k^i, η_k^i, μ_k^i, generated by algorithm* (37)–(38) *are such that with probability 1 for all $i \in V$:*

(a) The estimates x_k^i converge to the optimal point x^*.
(b) The estimates η_k^i converge to the optimal value η^* of the min-max problem.
(c) The dual iterates μ_k^i converge to a (random) optimal dual variable μ_i^*.

Proof. The proof proceeds by applying the Robbins-Siegmund result (Theorem 2) to the relation of Lemma 7. By our assumption on the step sizes we have $\sum_{k=0}^{\infty}\alpha_k^2 < \infty$, which immediately yields $\sum_{k=0}^{\infty}\alpha_k^2 m\frac{(C+v)^2}{2\sigma_x} < \infty$. Since, the projection step in algorithm (38) constrains the dual variables μ_k^i to lie in the interval $[0,1]$, it follows that $\sum_{k=0}^{\infty}\alpha_k^2\sum_{i=1}^{m}\frac{(1/m-\mu_k^i)^2}{2\sigma_\eta} < \infty$ with probability 1. Moreover, as the constraint set X and the set D are compact, and f_i are continuous, the term $(f_i(\tilde{x}_k^i)-\tilde{\eta}_k^i)^2$ is uniformly bounded for all i and k, implying that $\sum_{k=0}^{\infty}\alpha_k^2\sum_{i=1}^{m}\frac{(f_i(\tilde{x}_k^i)-\tilde{\eta}_k^i)^2}{2\sigma_{\mu,i}} < \infty$ with probability 1. From Lemma 6 we have $\sum_{k=0}^{\infty}\alpha_k\left(C\sum_{j=1}^{m}\|\hat{x}_k-x_k^j\|+|\hat{\eta}_k-\eta_k^j|\right) < \infty$ with probability 1. As $\sum_{k=0}^{\infty}\alpha_k=\infty$, it follows that with probability 1,

$$\lim_{k\to\infty}\sum_{j=1}^{m}\left(\|\hat{x}_k-x_k^j\|+|\hat{\eta}_k-\eta_k^j|\right)=0.$$

By the saddle-point Theorem 1 we have $\mathscr{L}(\hat{x}_k,\hat{\eta}_k,\mu^*)-\mathscr{L}(x^*,\eta^*,\mu_k)\geq 0$. Thus, all the conditions of Theorem 2 are satisfied.

By applying Theorem 2 we infer that $\mathbf{B}(z^*,\mu^*,\mathbf{z}_k,\mu_k)$ converges for every dual-optimal μ^* with probability 1, and that $\sum_{k=0}^{\infty}\alpha_k\left(\mathscr{L}(\hat{x}_k,\hat{\eta}_k,\mu^*)-L(x^*,\eta^*,\mu_k)\right)<\infty$ also holds with probability 1. Since $\sum_{k=0}^{\infty}\alpha_k=\infty$, it follows that with probability 1,

$$\lim_{k\to\infty}\left(\mathscr{L}(\hat{x}_k,\hat{\eta}_k,\mu^*)-L(x^*,\eta^*,\mu_k)\right)=0.$$

As the sequences $\{x_k^i\}$, $\eta_k^i\}$, $\{\mu_k^i\}$ are bounded, they have accumulation points in the sets X, D and $I=[0,1]$, respectively. We can use Cantor diagonalization-type argument to select a subsequence $\{k_\ell\}$ along which the following relations hold with probability 1:

$$\lim_{\ell\to\infty}(\hat{x}_{k_\ell},\hat{\eta}_{k_\ell},\mu_{k_\ell})=(\bar{x},\bar{\eta},\bar{\mu})\qquad\text{with}\quad(\bar{x},\bar{\eta},\bar{\mu})\in X\times D\times I^m,\tag{51}$$

$$\lim_{\ell\to\infty}\left(\mathscr{L}(\hat{x}_{k_\ell},\hat{\eta}_{k_\ell},\mu^*)-\mathscr{L}(x^*,\eta^*,\mu_{k_\ell})\right)=0,\tag{52}$$

$$\lim_{\ell\to\infty}\|\hat{x}_{k_\ell}-x_{k_\ell}^j\|=0,\qquad\lim_{\ell\to\infty}|\hat{\eta}_{k_\ell}-\eta_{k_\ell}^j|=0\qquad\text{for all } j\in V.\tag{53}$$

We now examine the consequences of these relations. Since (x^*,η^*,μ^*) is a saddle-point of the Lagrangian, it follows that $\mathscr{L}(\bar{x},\bar{\eta},\mu^*)\geq\mathscr{L}(x^*,\eta^*,\mu^*)\geq\mathscr{L}(x^*,\eta^*,\bar{\mu})$, implying that both inequalities hold as equalities. Therefore,

$$(\bar{x},\bar{\eta})\in\operatorname{argmin}_{(x,\eta)\in X\times D}\mathscr{L}(x,\eta,\mu^*),\qquad\bar{\mu}\in\operatorname{argmax}_{\mu\in I^m}\mathscr{L}(x^*,\eta^*,\mu).$$

Since (x^*, η^*) is the unique solution of problem (34) (by our assumption), the preceding relations together with the KKT conditions (Theorem 4) imply that $(\bar{x}, \bar{\eta}) = (x^*, \eta^*)$ and that $\bar{\mu}$ is an optimal dual multiplier.

It remains to show that the whole sequences converge to the desired optimal points. From (53) it follows that $\lim_{\ell \to \infty} \|x^j_{k_\ell} - x^*\| = 0$ and $\lim_{\ell \to \infty} |\eta^j_{k_\ell} - \eta^*| = 0$ with probability 1 for all j, implying $\lim_{\ell \to \infty} \mathbf{B}(z^*, \bar{\mu}, \mathbf{z}_{k_\ell}, \mu_{k_\ell}) = 0$. However, we have established that the sequence $\mathbf{B}(z^*, \mu^*, \mathbf{z}_k, \mu_k)$ converges with probability 1 for any optimal dual vector μ^*, so it converges with $\mu^* = \bar{\mu}$. Thus, it follows $\lim_{k \to \infty} \mathbf{B}(z^*, \bar{\mu}, \mathbf{z}_k, \mu_k) = 0$ with probability 1. Owing to the strong convexity of the Bregman functions we have

$$\mathbf{B}(z^*, \bar{\mu}, \mathbf{z}_k, \mu_k) \geq \sum_{i=1}^{m} \left(\frac{\sigma_x}{2} \|x^* - x^i_k\|^2 + \frac{\sigma_\eta}{2} |\eta^* - \eta^i_k|^2 + \frac{\sigma_{\mu,i}}{2} |\bar{\mu}_i - \mu^i_k|^2 \right),$$

which yields $x^i_k \to x^*$, $\eta^i_k \to \eta^*$ and $\mu^i_k \to \bar{\mu}_i$ with probability 1 for all $i \in V$. □

5 Min-Max Game Against Exogenous Player

In this section we consider a different formulation than the one discussed so far. We consider the case when the network of cooperative agents need to solve a min-max game against an exogenous player. The exogenous player is a malicious agent/nature which adversely affects the cost of each agent. Let us denote the action of the adversarial agent by ξ. We also require that the feasible set of allowable actions ξ is a compact set denoted by Θ. The objective is to solve:

$$\min_{x \in X} \max_{\xi \in \Theta} \sum_{i=1}^{m} f_i(x, \xi). \tag{54}$$

This can be thought of as the robust version of the problem considered in [35, 29], where the optimization problem of the form $\min_{x \in X} \sum_{i=1}^{m} \mathbb{E}_\xi \left[f_i(x, \xi) \right]$ was considered. In certain cases when it is desired to model the unknown signal ξ as lying in an uncertainty set Θ the robust version of problem (54) is more suitable. The problem (54) could alternatively be thought of as a zero sum game between the exogenous player and the network. To guarantee the existence of a min-max optimal solution to (54) we impose the following assumption.

Assumption 4. *Let the following hold:*

(a) The functions f_i are continuous over some open set containing $X \times \Theta$.
(b) The cost functions $f_i(x, \xi)$ are convex in x for every fixed value of $\xi \in \Theta$, and concave in ξ for every fixed $x \in X$.
(c) The constraint sets X and Θ are convex and compact.

Under Assumption 4 the min-max problem (54) admits a solution set $X^* \times \Theta^*$ such that for any $x^* \in X^*$ and $\xi^* \in \Theta^*$, we have the saddle-point property [4]:

$$\sum_{i=1}^{m} f_i(x^*,\xi) \leq \sum_{i=1}^{m} f_i(x^*,\xi^*) \leq \sum_{i=1}^{m} f_i(x,\xi^*) \quad \text{for all } x \in X \text{ and } \xi \in \Theta. \tag{55}$$

Let $B_x(\cdot,\cdot)$ and $B_\xi(\cdot,\cdot)$ be Bregman-distance functions for the sets X and Θ, respectively. We propose the following algorithm for min-max problem (54): at each iteration, at first, the agents update using the estimates x_k^j and ξ_k^i and obtain intermittent estimates

$$\begin{bmatrix} \tilde{x}_k^i \\ \tilde{\xi}_k^i \end{bmatrix} = \sum_{j=1}^{m} [W_k]_{ij} \begin{bmatrix} x_k^j \\ \xi_k^j \end{bmatrix}, \tag{56}$$

where W_k is a weight matrix as in (10). Using these intermittent adjustment, every agent updates according to the following rules:

$$\begin{aligned} x_{k+1}^i &= \text{argmin}_{y \in X} \left[\alpha_k \langle \nabla_x f_i(\tilde{x}_k^i, \tilde{\xi}_k^i), y \rangle + B_x(y, \tilde{x}_k^i) \right], \\ \xi_{k+1}^i &= \text{argmin}_{\zeta \in \Theta} \left[-\alpha_k \langle \nabla_\xi f_i(\tilde{x}_k^i, \tilde{\xi}_k^i), \zeta \rangle + B_\xi(\zeta, \tilde{\xi}_k^i) \right], \end{aligned} \tag{57}$$

where ∇_x and ∇_ξ denote the partial derivative operators with respect to the variables x and ξ, respectively. The initial points x_0^i and ξ_0^i satisfy $x_0^i \in X$ and $\xi_0^i \in \Theta$ for all i. It is also assumed that the constraint sets X and Θ are common knowledge for all agents $i \in V$.

The analysis of the algorithm follows along lines similar to that of the primal-dual algorithm (37)–(38). This can be seen in light of the fact that the primal-dual algorithm computes a saddle-point of the Lagrangian function in (5), whereas algorithm (56)–(57) computes a saddle-point of problem (54). A major difference between the algorithms is the fact that in (38) the agents update their own local dual variables μ_k^i, whereas in the algorithm (57) the agents update the whole vector ξ which is coupling the agents. Note that there is no stochasticity in the current formulation unless we consider a stochastic model of the network. The final result is that asymptotically the agents estimates x_k^i and ξ_k^i converge to a common min-max optimal pair (x^*,ξ^*). We formalize the statement in the following theorem.

Theorem 6. *Let Assumption 1 and 4 hold. Assume that problem* (54) *has an optimal set of the form* $\{x^*\} \times \Theta^*$*. Moreover, assume that the Bregman-distance functions* $B_x(y,v)$ *and* $B_\xi(\zeta,\phi)$ *are convex in their second arguments* v *and* ϕ*, respectively for every fixed* y *and* ζ*. If the step sizes* α_k *in algorithm* (56)–(57) *are chosen to satisfy* $\sum_{k=0}^{\infty} \alpha_k = \infty$ *and* $\sum_{k=0}^{\infty} \alpha_k^2 < \infty$*, then the local variables* (x_k^i, ξ_k^i) *converge to a common saddle-point solution* (x^*,ξ^*) *of the min-max problem* (54)*, for all* $i \in V$.

Proof. The proof is similar to the proof of Theorem 5, with the Lyapunov function $\sum_{i=1}^{m} \left(B_x(x^*, x_k^i) + B_\xi(\xi^*, \xi_k^i) \right)$ for an arbitrary saddle-point solution (x^*,ξ^*). □

6 Uplink Power Control

In this section we show the suitability of our algorithms in (10)–(11) and (37)–(38) to achieve a min-max fair allocation of utility in a cellular network. We will keep our discussion brief and refer the readers to [12] for a general discussion on the power allocation problem. We will be using the formulation discussed in [30].

There are m mobile users (MU) in neighboring cells communicating with their respective base stations (BS) using a common wireless channel. Let p_i denote the power used by MU i to communicate with its base station. Due to the shared nature of the wireless medium the total received SINR at BS i is given by

$$\gamma_i(\bar{p},\bar{h}_i) = \frac{p_i h_{i,i}^2}{\sigma_i^2 + \sum_{j\neq i} p_j h_{i,j}^2},$$

where $h_{i,j}$ is the channel coefficient between MU j and BS i, and σ_i^2 is the receiver noise variance. The vector containing power variables p_i is denoted $\bar{p}$ and the vector of channel coefficients at BS i is denoted $\bar{h}_i$. The power variables are non-negative and constrained to a maximum value of p_t, i.e., $0 \leq p_i \leq p_t$ for all i.

Let $U_i(\gamma_i(\bar{p},\bar{h}_i))$ be the utility derived by BS i and $V(p_i)$ be a cost function penalizing excessive power. We are interested in finding an allocation that minimizes the worst case loss to any agent i, which amounts to solving the following problem:

$$\min_{\bar{p}\in\Pi} \max_{i\in V} \left[V(p_i) - U_i(\gamma_i(\bar{p},\bar{h}_i))\right],$$

where $\Pi = \{\bar{p} \in \mathbb{R}^m \mid 0 \leq p_i \leq p_t \text{ for all } i\}$ and p_t is the maximum power. We consider the logarithmic utility function $U_i(u) = \log(u)$ for $u > 0$. Using the transformation $p_i = e^{x_i}$, it can be shown that the preceding problem can be cast in the form of (1), with a cost function for each base station i given by:

$$f_i(x) = \log\left(\sigma_i^2 h_{i,i}^{-2} e^{-x_i} + \sum_{j\neq i} h_{i,i}^{-2} h_{j,i}^2 e^{x_j - x_i}\right) + V(e^{x_i}),$$

and $X = \{x \in \mathbb{R}^m \mid x_i \leq \log(p_t) \text{ for all } i\}$.

We have considered a cellular network of 16 square cells of the same size ($m = 16$). The connectivity network for the BSs is shown in Figure 1. Within each cell, the MU is randomly located (with a uniform distribution over the cell) and the base station is located at the center of the cell. The channel coefficient $h_{i,j}$ is assumed to decay as the fourth power of the distance between the MU j and the BS i. The shadow fading is assumed to be log-normal with variance 0.1. The receiver noise variance σ_i^2 is taken to be 0.01. The cost of the power is modeled as $V(p_i) = 10^{-3} p_i$.

In the simulations, there are no stochastic errors, i.e., all gradients and subgradients are evaluated without stochastic errors. Four algorithms are used, namely, the standard (centralized) gradient descent algorithm (applied to the penalized problem), a centralized primal-dual algorithm (that computes a saddle point of the min-max problem), the distributed exact penalty (10)–(11), and the distributed

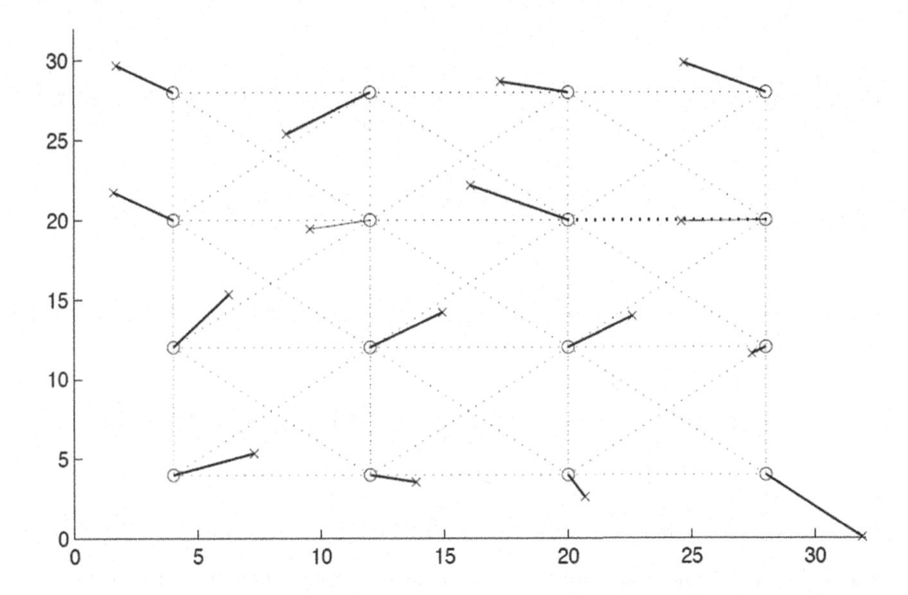

Fig. 1 The circles denote the BSs. The dotted lines denote the communication links between adjacent BSs. The crosses denote the MUs. The bold lines connect each MU to its respective BS.

primal-dual (37)–(38). The standard (centralized) gradient descent algorithm is used to determine the optimal min-max allocation, while the centralized primal-dual is used to get a sense of dual optimal variables. In the distributed algorithms, the weight matrices W_k are all equal since the connectivity graph is static, i.e., $W_k = W$ for all k. The weights W_{ij} are given by:

$$W_{ij} = \frac{1}{\max\{|N_i|, |N_j|\}} \quad \text{if } i \text{ and } j \text{ are neighbors,}$$

and otherwise $W_{ij} = 0$, where $|N_i|$ denotes the cardinality of the neighbor set N_i (which includes agent i itself). The stepsize is $\alpha_k = \frac{10}{k}$ in the centralized methods, $\alpha_k = \frac{50}{k^{0.65}}$ in the distributed exact penalty method, and $\alpha_k = \frac{4}{k^{0.6}}$ in the distributed primal-dual method. The penalty parameter r_i is 1.3 for all i in algorithm (10)–(11). The Bregman-distance generating functions are the Euclidean norms. Each algorithm is run for 4000 iterations.

Figure 2 shows the behavior of algorithms (10)–(11) and (37)–(38). As seen in the figure, both centralized algorithms perform the best (as they have the whole knowledge of the problem information) and they have a similar behavior. The distributed algorithms are slightly worse than the centralized, which is expected due to their "decentralized" incomplete knowledge of the problem. Of the two distributed algorithms, the primal-dual algorithm is worse, as it often assigns much larger allocations than the distributed exact penalty method. Primal-dual algorithms (in absence

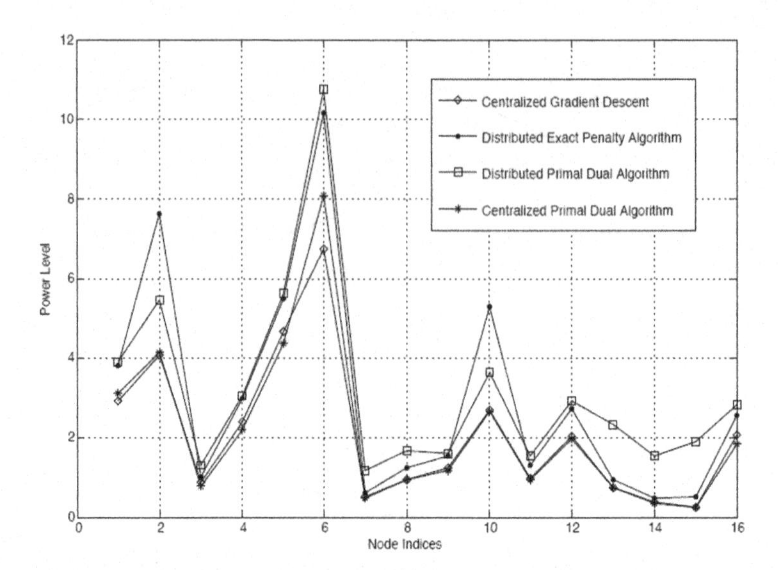

Fig. 2 The final iterate values after 4000 iterations of the algorithms. The plot shows the allocations achieved by Centralized Gradient Descent, Distributed Exact Penalty Algorithm, Distributed Primal-Dual Algorithm, and Centralized Primal-Dual Algorithm.

of strong convexity of the primal) are known to be highly sensitive to the choices of dual variables, which may be reflected in the results we see in Figure 2, where the primal-dual algorithms use the initial values $\mu_i = \frac{1}{m}$ for the multipliers.

7 Conclusion

We presented distributed algorithms for solving stochastic min-max optimization problems in networks. We developed two algorithms based on Bregman-distance functions. The first algorithm uses a non-differentiable penalty function to translate the min-max problem to a format which is suitable for distributed algorithms. The second algorithm is based on the primal-dual iterative update scheme. In both of these algorithms we allow the presence of stochastic subgradient noise. We provided conditions on the dynamic network under which we can guarantee almost sure convergent behavior of the algorithms. We illustrated the applicability of the algorithms on a power allocation problem in a cellular network.

The effectiveness of these algorithms is highly dependable on the underlying connectivity structure of the agent network. For future work, we plan to investigate error bounds for the proposed algorithms, which will capture the scalability of the algorithms with the number m of agents. Based on our prior work [31], we know that these algorithms can scale at best in the order of $m^{3/2}$ when the sum of the functions f_i is to be minimized and the $\sum_{i=1}^{m} f_i$ is strongly convex. We believe that this bound is also achievable by the proposed algorithms for a class of functions

f_i, such as linear for example. Such results will also provide better insights into the practical short-term behavior of these algorithms.

Acknowledgements. This work has been supported by the NSF Grant CMMI-0742538.

References

1. Agarwal, A., Duchi, J., Wainwright, M.: Dual averaging for distributed optimization: Convergence analysis and network scaling. IEEE Transactions on Automatic Control (2011) (to appear)
2. Arrow, K.J., Hurwicz, L., Uzawa, H.: Studies in Linear and Non-Linear Programming. Stanford University Press, Stanford (1958)
3. Bertsekas, D.P.: Necessary and sufficient conditions for a penalty method to be exact. Mathematical Programming 9, 87–99 (1975)
4. Bertsekas, D.P., Nedić, A., Ozdaglar, A.: Convex Analysis and Optimization. Athena Scientific, Belmont (2003)
5. Bertsekas, D.P., Tsitsiklis, J.N.: Parallel and distributed computation: numerical methods. Prentice-Hall, Inc., Upper Saddle River (1989)
6. Bertsekas, D.P., Tsitsiklis, J.N.: Gradient convergence in gradient methods with errors. Siam. J. Optim. 10(3), 627–642 (2000)
7. Billingsley, P.: Probability and Measure. John Wiley and Sons (1979)
8. Boche, H., Wiczanowski, M., Stanczak, S.: Unifying view on min-max fairness and utility optimization in cellular networks. In: 2005 IEEE Wireless Communications and Networking Conference, vol. 3, pp. 1280–1285 (March 2005)
9. Borkar, V.S.: Stochastic Approximation: A Dynamical Systems Viewpoint. Cambridge University Press (2008)
10. Bregman, L.: The relaxation method of finding the common point of convex sets and its application to the solution of problems in convex programming. USSR Computational Mathematics and Mathematical Physics 7(3), 200–217 (1967)
11. Censor, Y.A., Zenios, S.A.: Parallel Optimization: Theory, Algorithms and Applications. Oxford University Press (1997)
12. Chiang, M., Hande, P., Lan, T., Tan, W.C.: Power Control in Wireless Cellular Networks. Found. Trends Netw. 2(4), 381–533 (2008)
13. Ermoliev, Y.: Stochastic quasi-gradient methods and their application to system optimization. Stochastics 9(1), 1–36 (1983)
14. Ermoliev, Y.: Stochastic quazigradient methods. In: Numerical Techniques for Stochastic Optimization, pp. 141–186. Springer, Heidelberg (1988)
15. Hastie, T., Tibshirani, R., Friedman, J.H.: The Elements of Statistical Learning. Springer (2003)
16. Jadbabaie, A., Lin, J., Morse, S.: Coordination of groups of mobile autonomous agents using nearest neighbor rules. IEEE Transactions on Automatic Control 48, 988–1001 (2003)
17. Johansson, B., Rabi, M., Johansson, M.: A randomized incremental subgradient method for distributed optimization in networked systems. SIAM Journal on Optimization 20(3), 1157–1170 (2009)
18. Kar, S., Moura, J.M.F.: Distributed consensus algorithms in sensor networks with imperfect communication: link failures and channel noise. IEEE Tran. Signal Process. 57(1), 355–369 (2009)

19. Lobel, I., Ozdaglar, A.: Distributed subgradient methods for convex optimization over random networks. IEEE Transactions on Automatic Control 56(6), 1291–1306 (2011)
20. Mosk-Aoyama, D., Roughgarden, T., Shah, D.: Fully Distributed Algorithms for Convex Optimization Problems. In: Pelc, A. (ed.) DISC 2007. LNCS, vol. 4731, pp. 492–493. Springer, Heidelberg (2007)
21. Nedić, A.: Asynchronous broadcast-based convex optimization over a network. IEEE Transactions on Automatic Control 56(6), 1337–1351 (2011)
22. Nedić, A., Olshevsky, A., Ozdaglar, A., Tsitsiklis, J.N.: Distributed subgradient algorithms and quantization effects. In: Proceedings of the 47th IEEE Conference on Decision and Control, pp. 4177–4184 (2008)
23. Nedić, A., Ozdaglar, A.: Distributed subgradient methods for multi-agent optimization. IEEE Transactions on Automatic Control 54(1), 48–61 (2009)
24. Nedić, A., Ozdaglar, A.: Subgradient methods for saddle-point problems. Journal of Optimization Theory and Applications 142(1), 205–228 (2009)
25. Nedić, A., Ozdaglar, A., Parrilo, P.A.: Constrained consensus and optimization in multi-agent networks. IEEE Transactions on Automatic Control 55, 922–938 (2010)
26. Nemirovski, A., Juditsky, A., Lan, G., Shapiro, A.: Robust stochastic approximation approach to stochastic programming. SIAM J. on Optimization 19(4), 1574–1609 (2008)
27. Polyak, B.T.: Introduction to Optimization. Optimization Software, Inc., New York (1987)
28. Rabbat, M., Nowak, R.D.: Distributed optimization in sensor networks. In: IPSN, pp. 20–27 (2004)
29. Ram, S.S., Nedić, A., Veeravalli, V.V.: Distributed stochastic subgradient projection algorithms for convex optimization. Journal of Optimization Theory and Applications 147(3), 516–545 (2010)
30. Ram, S.S., Veeravalli, V.V., Nedić, A.: Distributed non-autonomous power control through distributed convex optimization. In: IEEE INFOCOM, pp. 3001–3005 (2009)
31. Sundhar Ram, S., Nedić, A., Veeravalli, V.V.: Asynchronous gossip algorithms for stochastic optimization: Constant stepsize analysis. In: Diehl, M., Glineur, F., Jarlebring, E., Michiels, W. (eds.) Recent Advances in Optimization and its Applications in Engineering, 14th Belgian-French-German Conference on Optimization (BFG), pp. 51–60 (2010)
32. Ram, S.S., Nedić, A., Veeravalli, V.V.: A new class of distributed optimization algorithms: Application to regression of distributed data. Optimization Methods and Software 27(1), 71–88 (2012)
33. Robbins, H., Siegmund, D.: A convergence theorem for nonnegative almost supermartingales and some applications. In: Rustagi, J.S. (ed.) Proceedings of a Symposium on Optimizing Methods in Statistics, pp. 233–257. Academic Press, New York (1971)
34. Srikant, R.: The Mathematics of Internet Congestion Control. Birkhäuser, Boston (2003)
35. Srivastava, K., Nedić, A.: Distributed asynchronous constrained stochastic optimization. IEEE Journal of Selected Topics in Signal Processing 5(4), 772–790 (2011)
36. Srivastava, K., Nedić, A., Stipanović, D.: Distributed min-max optimization in networks. In: 17th International Conference on Digital Signal Processing (2011)
37. Stanković, S.S., Stanković, M.S., Stipanović, D.M.: Decentralized parameter estimation by consensus based stochastic approximation. In: 2007 46th IEEE Conference on Decision and Control, pp. 1535–1540 (December 2007)
38. Tsitsiklis, J.N.: Problems in decentralized decision making and computation. PhD thesis, Massachusetts Institute of Technology, Boston (1984)
39. Tsitsiklis, J.N., Athans, M.: Convergence and asymptotic agreement in distributed decision problems. IEEE Trans. Automat. Control 29, 42–50 (1984)

A Probability Collectives Approach for Multi-Agent Distributed and Cooperative Optimization with Tolerance for Agent Failure

Anand J. Kulkarni and Kang Tai*

Abstract. Centralized systems are vulnerable to single point failures that may severely affect their performance. On the other hand, in the case of a distributed and decentralized algorithm, the system is more robust as it is controlled by its autonomous subsystems or agents. This chapter intends to demonstrate the inherent ability of a distributed and decentralized agent-based optimization technique referred to as Probability Collectives (PC) to accommodate agent failures. The approach of PC is a framework for optimization of complex systems by decomposing them into smaller subsystems to be further treated in a distributed and decentralized way. The system can be viewed as a Multi-Agent System (MAS) with rational and self-interested agents optimizing their local goals. At the core of the PC optimization methodology are the concepts of Deterministic Annealing in Statistical Physics, Game Theory and Nash Equilibrium. A specially developed Circle Packing Problem (CPP) with a known true optimum solution will be solved to demonstrate the ability of the PC approach to tolerate instances of agent failure. The strengths, weaknesses and future research directions of the PC methodology will also be discussed.

1 Introduction

The decomposition of an entire system into smaller subsystems and optimizing them in a distributed and decentralized way to reach the system level optimum is one of the emerging approaches to deal with the growing complexity and uncertainty encountered in real world problems. These subsystems together can be

---

Anand J. Kulkarni · Kang Tai
School of Mechanical and Aerospace Engineering, Nanyang Technological University, 50 Nanyang Avenue, Singapore 639798, Singapore
e-mail: {kulk0003,mktai}@ntu.edu.sg

* Corresponding author.

I. Czarnowski et al. (Eds.): Agent-Based Optimization, SCI 456, pp. 175–201.
DOI: 10.1007/978-3-642-34097-0_8

viewed as a collective, which in other words is a group of learning agents or a Multi-Agent System (MAS) [1-5]. In a distributed MAS, the rational and self-interested behavior of the agents is very important to achieve the best possible local goal/reward/payoff, but it is not trivial to make such agents work collectively to achieve the best possible global or system objective. Certainly, the major advantage of the distributed and decentralized optimization approach is its immunity to single point failure. The centralized system is vulnerable to single point failures and its performance may be severely affected if an agent in the system fails. On the other hand, in the case of a distributed and decentralized algorithm, the system is more robust as the system is controlled by its autonomous subsystems. The immunity to agent failure is essential in numerous applications including UAV formation and collaborative path planning [3], urban traffic control [6-11], sensor networks [12-18], etc.

Probability Collectives (PC) in the framework of Collective Intelligence (COIN) was first proposed by Dr. David Wolpert in 1999 [19]. It is an emerging distributed optimization methodology for modelling and controlling distributed MAS, inspired from a sociophysics viewpoint with deep connections to Game Theory, Statistical Physics, and Optimization [2, 3, 19, 20]. PC considers every variable in the system as an independent agent. These agents assign probability distributions to its possible set of actions/moves. Every agent independently updates the probability distribution affecting its local goal which in turn also affects the global or system objective [2, 3, 19, 20]. The process continues and reaches equilibrium when no further increase in reward is possible for the individual agent by changing its actions further. This equilibrium concept is referred to as Nash equilibrium [21]. It is important to mention that the approach works on probability distributions and thereby directly incorporates uncertainty. The approach of PC has been implemented for solving both unconstrained [3, 4, 5, 12-20, 22, 23, 24, 25] as well as constrained [1, 3, 26, 27, 31, 32] optimization problems. A concise summary of the PC literature is discussed below.

According to [24], PC outperformed Genetic Algorithms (GAs) in terms of robustness, reproducibility, rate of descent, trapping in false minima and long term optimization when solving benchmark problems such as Schaffer's function, Rosenbrock function, Ackley Path function and Michalewicz Epistatic function. A variation of the original PC approach in [2, 3, 27, 28] referred to as Sequentially Updated PC (SPC) [25] performed better with higher dimension Hartman's functions only but failed to converge in the target assignment game. The decentralized PC architecture also outperformed its centralized architecture solving the 8-queens problem [22]. Two different PC approaches were proposed in [26] avoiding airplanes collision. In the first approach, every airplane was assumed to be an autonomous agent selecting their individual paths to avoid collision with other airplanes travelling in the neighbourhood. In the semi centralized approach, every airplane was given a chance to become a host airplane which computed and distributed the solution to all other airplanes.

The approach of PC [27, 28] was also tested on the discrete constrained problem of university course scheduling [31], but the implementation failed to generate any feasible solution. PC was also tested solving the discrete constrained problem of optimizing the cross-sections of individual bars and segments of a 10 bar truss

[28]. The approach of PC was successfully applied solving complex combinatorial optimization problem of airplane fleet assignment having the goal of minimization of the number of flights with 129 variables and 184 constraints [27]. This divided the communication and computational load into agents and also latency in the system which could have resulted in the growing possibility of conflict in schedules and continuity. The approach of PC was also successfully applied solving combinatorial optimization problems such as the joint optimization of the routing and resource allocation in wireless networks [12-18].

The authors of this chapter also tested the approach of PC for solving continuous unconstrained segmented beam problem [4], with the sampling method as well as associated sampling space updating scheme of the original PC approach modified. This modified PC approach was also validated successfully by optimizing the Rosenbrock function [5]. It was also applied for solving two test cases of the NP-hard combinatorial problem of Multi-Depot Multiple Travelling Salesmen Problem (MDMTSP) [1] as well as the cases of Single Depot MTSP (SDMTSP) [29]. Moreover, a constrained PC approach using a penalty function method was developed and applied to three test problems in [30]. In addition, in order to make PC an even more versatile optimization algorithm, a variation of the feasibility-based rule originally proposed in [33] and further implemented in [35-40] was employed for solving two cases of the Circle Packing Problem (CPP) [32]. It is worth to mention here that both the cases yielded the true optimum solution in every run of the PC, which clearly demonstrated its ability to avoid the tragedy of commons. A more generic variation of the feasibility-based rule was successfully implemented for solving three variations of the Sensor Network Coverage Problem [33].

The above discussion shows that PC is versatile and applicable to variegated areas including constrained optimization problems; however, PC has never been tested for the practically important agent failure case. This chapter attempts to demonstrate the capability of PC as a distributed optimization approach to deal with an agent failure scenario. The solution highlights its potential to deal with agent failures which may arise in real world complex problems including urban traffic control, formation of airplanes fleet and mid-air collision avoidance, etc.

The remainder of this chapter is organized as follows. Section 2 provides the framework and detailed formulation of the constrained PC method. It includes the formulation of the homotopy function, the constraint handling technique using the feasibility-based rule and the concept of Nash equilibrium. In Section 3, the constrained PC approach addressing the agent failure scenario is demonstrated by solving the CPP. It also includes an associated problem specific heuristic technique. The evident features, advantages, limitations and conclusions of the presented work and associated future directions are discussed in Section 5.

2 The Constrained PC Framework and Formulation

The variables in an optimization problem are considered by PC to be individual self interested learning agents/players of a game being played iteratively [2, 3, 27, 28]. While working in some definite direction, these agents select actions over a particular interval and receive some local rewards on the basis of the system objective achieved because of those actions. In other words, these agents optimize their local rewards or payoffs which at the same time also optimize the system level performance. The process iterates and reaches equilibrium (referred to as Nash equilibrium) when no further increase in the reward is possible for the individual agent through changing its actions further. Moreover, the method of PC theory is an efficient way of sampling the joint probability space, converting the problem into the convex space of probability distribution. PC allocates probability values to each agent's moves, and hence directly incorporates uncertainty. This is based on prior knowledge of the recent action or behavior selected by all other agents. In short, the agents in the PC framework need to have knowledge of the environment along with every other agent's recent action or behavior.

At every iteration, each agent randomly samples the moves/strategies from within its own sampling set (i.e. its own sampling interval) as well as from within other agents' strategy sets and computes the corresponding system objectives. The other agents' strategy sets are modeled by each agent based on their recent actions or behavior only, i.e. based on partial knowledge. By minimizing the collection of system objectives, every agent identifies the possible strategy which contributes the most towards the minimization of the collection of system objectives. Such a collection of functions is computationally expensive to minimize and also may lead to local minima [3]. In order to avoid this difficulty, the collection of system objectives is deformed into another topological space forming the homotopy function parameterized by computational temperature T [41-44]. Due to its analogy to Helmholtz free energy [20, 41-45], the approach of Deterministic Annealing (DA) converting the discrete variable space into continuous variable space of probability distribution is applied in minimizing the homotopy function. At every successive temperature drop, the minimization of the homotopy function is carried out using a second order optimization scheme such as the Nearest Newton Descent Scheme [1-5, 12-18, 22-31] or BFGS Scheme [32, 33], etc.

At the end of each iteration, each agent i converges to a probability distribution clearly distinguishing the contribution of its every corresponding strategy value. For every agent, the strategy value with the maximum probability value is referred to as the favorable strategy and is used to compute the system objective and corresponding constraint functions. This system objective and corresponding strategy values are accepted based on a variation of the feasibility-based rule defined in [34] and further successfully implemented in [35-40]. This rule allows the objective function and the constraint information to be considered separately. The rule can be described as follows:

(a) Any feasible solution is preferred over any infeasible solution
(b) Between two feasible solutions, the one with better objective is preferred
(c) Between two infeasible solutions, the one with fewer constraint violations is preferred.

In addition to the above, a perturbation approach is also incorporated to avoid premature convergence. It perturbs the individual agent's favorable strategy set based on its reciprocal and associated predefined interval. The solution is accepted if the feasibility is maintained. In this way, the algorithm continues until convergence by selecting the samples from the neighborhood of the recent favorable strategies. The neighborhood space is reduced or expanded according to the improvement in the system objective for a predefined number of iterations.

In some of the applications, the agents are also needed to provide the knowledge of the inter-agent-relationship. It is one of the information/strategy sets which every other entitled agent is supposed to know. There is also global information that every agent is supposed to know. This allows agents to know the right to model other agents' actions or behavior. All of the decisions are taken autonomously by each agent considering the available information in order to optimize the local goals and hence to achieve the optimum global goal or system objective. The following section discusses the constrained PC procedure in detail.

2.1 Constrained PC Algorithm

Consider a general constrained problem (in the minimization sense) as follows:

$$\text{Minimize} \quad G \tag{1}$$

Subject to

s number of inequality constraints $g_j \leq 0, \quad j = 1, 2, ..., s$

w number of equality constraints $h_j = 0, \quad j = 1, 2, ..., w$

where the objective function G can take real and/or discrete variables. According to [46-48], the equality constraint $h_j = 0$ can be transformed into a pair of inequality constraints using a tolerance value δ as follows:

$$h_j = 0 \quad \Rightarrow \begin{cases} g_{s+j} = h_j - \delta \leq 0 & j = 1, 2, ..., w \\ g_{s+w+j} = -\delta - h_j \leq 0 \end{cases} \tag{2}$$

Thus, w equality constraints are replaced by $2w$ inequality constraints with the total number of constraints given by $t = s + 2w$. Then a generalized representation of the problem in equation (1) can be stated as follows:

$$\text{Minimize} \quad G \tag{3}$$

$$\text{Subject to} \quad g_j \leq 0, \quad j = 1, 2, ..., t$$

In the context of PC, the variables of the problem are considered as computational agents/players of a social game being played iteratively [3, 19]. Each agent i is given a predefined sampling interval referred to as $\Psi_i \in \left[\Psi_i^{lower}, \Psi_i^{upper}\right]$. As a general case, the interval can also be referred to as the sampling space. The lower limit Ψ_i^{lower} and upper limit Ψ_i^{upper} of the interval Ψ_i may be updated iteratively as the algorithm progresses.

Each agent i randomly samples $X_i^{[r]}$, $r = 1,2,...,m_i$ strategies from within the corresponding sampling interval Ψ_i forming a strategy set $\mathbf{X}_i$ represented as

$$\mathbf{X}_i = \{X_i^{[1]}, X_i^{[2]}, X_i^{[3]}, ..., X_i^{[m_i]}\} \quad , \quad i = 1,2,...,N \tag{4}$$

where N is the number of variables in the problem.

Every agent is assumed to have an equal number of strategies, i.e. $m_1 = m_2 = ... = m_i = ... = m_{N-1} = m_N$. The procedure of modified PC theory is explained below in detail with the algorithm flowchart in Figure 1.

The procedure begins with the initialization of the sampling interval Ψ_i for each agent i, temperature $T >> 0$ or $T = T_{initial}$ or $T \to \infty$ (simply high enough), the temperature step size α_T $(0 < \alpha_T \le 1)$, convergence parameter $\varepsilon = 0.0001$, algorithm iteration counter $n = 1$ and number of test iterations n_{test}. The value of α_T and n_{test} are chosen based on preliminary trials of the algorithm. Furthermore, the constraint violation tolerance μ is initialized to the number of constraints $|\mathbf{C}|$, i.e. $\mu = |\mathbf{C}|$ where $|\mathbf{C}|$ refers to the cardinality of the constraint vector $\mathbf{C} = [g_1, g_2, ..., g_t]$.

Step 1. Agent i selects its first strategy $X_i^{[1]}$ and samples randomly from other agents' strategies as well. This is a random guess by agent i about which strategies have been chosen by the other agents. This forms a 'combined strategy set' $\mathbf{Y}_i^{[1]}$ given by

$$\mathbf{Y}_i^{[1]} = \left\{X_1^{[?]}, X_2^{[?]}, ..., X_i^{[1]}, ..., X_{N-1}^{[?]}, X_N^{[?]}\right\} \tag{5}$$

The superscript [?] indicates that it is a 'random guess' and not known in advance. In addition, agent i forms one combined strategy set for every strategy r of its strategy set $\mathbf{X}_i$, as shown below.

$$\begin{aligned}
\mathbf{Y}_i^{[2]} &= \left\{X_1^{[?]}, X_2^{[?]}, ..., X_i^{[2]}, ..., X_{N-1}^{[?]}, X_N^{[?]}\right\} \\
\mathbf{Y}_i^{[3]} &= \left\{X_1^{[?]}, X_2^{[?]}, ..., X_i^{[3]}, ..., X_{N-1}^{[?]}, X_N^{[?]}\right\} \\
&\vdots \\
\mathbf{Y}_i^{[r]} &= \left\{X_1^{[?]}, X_2^{[?]}, ..., X_i^{[r]}, ..., X_{N-1}^{[?]}, X_N^{[?]}\right\} \\
&\vdots \\
\mathbf{Y}_i^{[m_i]} &= \left\{X_1^{[?]}, X_2^{[?]}, ..., X_i^{[m_i]}, ..., X_{N-1}^{[?]}, X_N^{[?]}\right\}
\end{aligned} \tag{6}$$

Similarly, all the remaining agents form their combined strategy sets.

Furthermore, every agent i computes m_i associated objective function values as follows:

$$\left[G\left(\mathbf{Y}_i^{[1]}\right), G\left(\mathbf{Y}_i^{[2]}\right), \ldots, G\left(\mathbf{Y}_i^{[r]}\right), \ldots, G\left(\mathbf{Y}_i^{[m_i]}\right)\right] \quad (7)$$

The ultimate goal of every agent i is to identify its strategy value which contributes the most towards the minimization of the sum of these system objective values, i.e. $\sum_{r=1}^{m_i} G(\mathbf{Y}_i^{[r]})$, hereafter referred to as the collection of system objectives.

Step 2. The minimum of the function $\sum_{r=1}^{m_i} G(\mathbf{Y}_i^{[r]})$ is very hard to achieve as the function may have many possible local minima. Moreover, directly minimizing this function is quite cumbersome as it may need excessive computational effort [3]. One of the ways to deal with this difficulty is to deform the function into another topological space by constructing a related and 'easier' function $f\left(\mathbf{X}_i\right)$. Such a method is referred to as the homotopy method [41-44]. The function $f\left(\mathbf{X}_i\right)$ can be referred to as 'easier' because it is easy to compute, the (global) minimum of such a function is known and easy to locate [41-46]. The deformed function can also be referred to as homotopy function J parameterized by computational temperature T represented as follows:

$$J_i\left(\mathbf{X}_i, T\right) = \sum_{r=1}^{m_i} G(\mathbf{Y}_i^{[r]}) - T f\left(\mathbf{X}_i\right) \quad , \quad T \in [0, \infty) \quad (8)$$

The approach of Deterministic Annealing (DA) is applied to minimize the homotopy function in equation (8). The motivation behind this is its analogy to the Helmholtz free energy [29, 30, 32, 33]. It suggests the conversion of discrete variables into random real valued variables such as probabilities. This converts the original collection of system objectives $\sum_{r=1}^{m_i} G(\mathbf{Y}_i^{[r]})$ into the 'expected collection of system objectives $\sum_{r=1}^{m_i} E\left(G(\mathbf{Y}_i^{[r]})\right)$. Furthermore, a suitable function for $f\left(\mathbf{X}_i\right)$ is chosen. The general choice is to use the entropy function $\mathrm{S}_i = -\sum_{r=1}^{m_i} q(X_i^{[r]}) \log_2 q(X_i^{[r]})$ [41-43].

homotopy function to the Helmholtz free energy discussed in [29, 30, 32, 33].

(a) Agent i assigns uniform probabilities to its strategies. This is because, at the beginning, the least information is available (the largest uncertainty and highest entropy) about which strategy is favorable for the minimization of the collection of system objectives $\sum_{r=1}^{m_i} G(\mathbf{Y}_i^{[r]})$. Therefore, at the beginning of the 'game', each agent's every strategy has probability $1/m_i$ of being most favorable. Therefore, probability of strategy r of agent i is

$$q(X_i^{[r]}) = 1/m_i \quad , \quad r = 1,2,...,m_i \tag{9}$$

Each agent i, from its every combined strategy set $\mathbf{Y}_i^{[r]}$ and corresponding system objective $G(\mathbf{Y}_i^{[r]})$ computed previously, further computes m_i corresponding expected system objective values $E\left(G(\mathbf{Y}_i^{[r]})\right)$ as follows 2, 3, 19, 20, 27, 28]:

$$E\left(G(\mathbf{Y}_i^{[r]})\right) = G(\mathbf{Y}_i^{[r]})\, q(X_i^{[r]}) \prod_{(i)} q(X_{(i)}^{[?]}) \tag{10}$$

where (i) represents every agent other than i. Every agent i then computes the expected collection of system objectives denoted by $\sum_{r=1}^{m_i} E\left(G(\mathbf{Y}_i^{[r]})\right)$. This also means that the PC approach can convert any discrete variables into continuous variable values in the form of probabilities corresponding to these discrete variables. As mentioned earlier, the problem now becomes continuous but still not easier to solve.

(b) Thus the homotopy function to be minimized by each agent i in equation (8) is modified as follows:

$$J_i\left(q(\mathbf{X}_i),T\right) = \sum_{r=1}^{m_i} E\left(G(\mathbf{Y}_i^{[r]})\right) - T\,S_i \tag{11}$$

where $T \in [0,\infty)$. When the temperature T is high enough, the entropy term dominates the expected collection of system objectives and the problem becomes very easy to be solved.

The following steps of DA are formulated based on the analogy of the

Step 3. In the author's previous work [1-3, 29, 30], Nearest Newton Descent Scheme [4, 5, 12-18, 22-28, 31] was implemented for minimizing the homotopy function $J_i\left(q(\mathbf{X}_i),T\right)$. Motivated from this scheme, the minimization of the homotopy function $J_i\left(q(\mathbf{X}_i),T\right)$ given in equation (11) is carried out using a suitable second order optimization technique

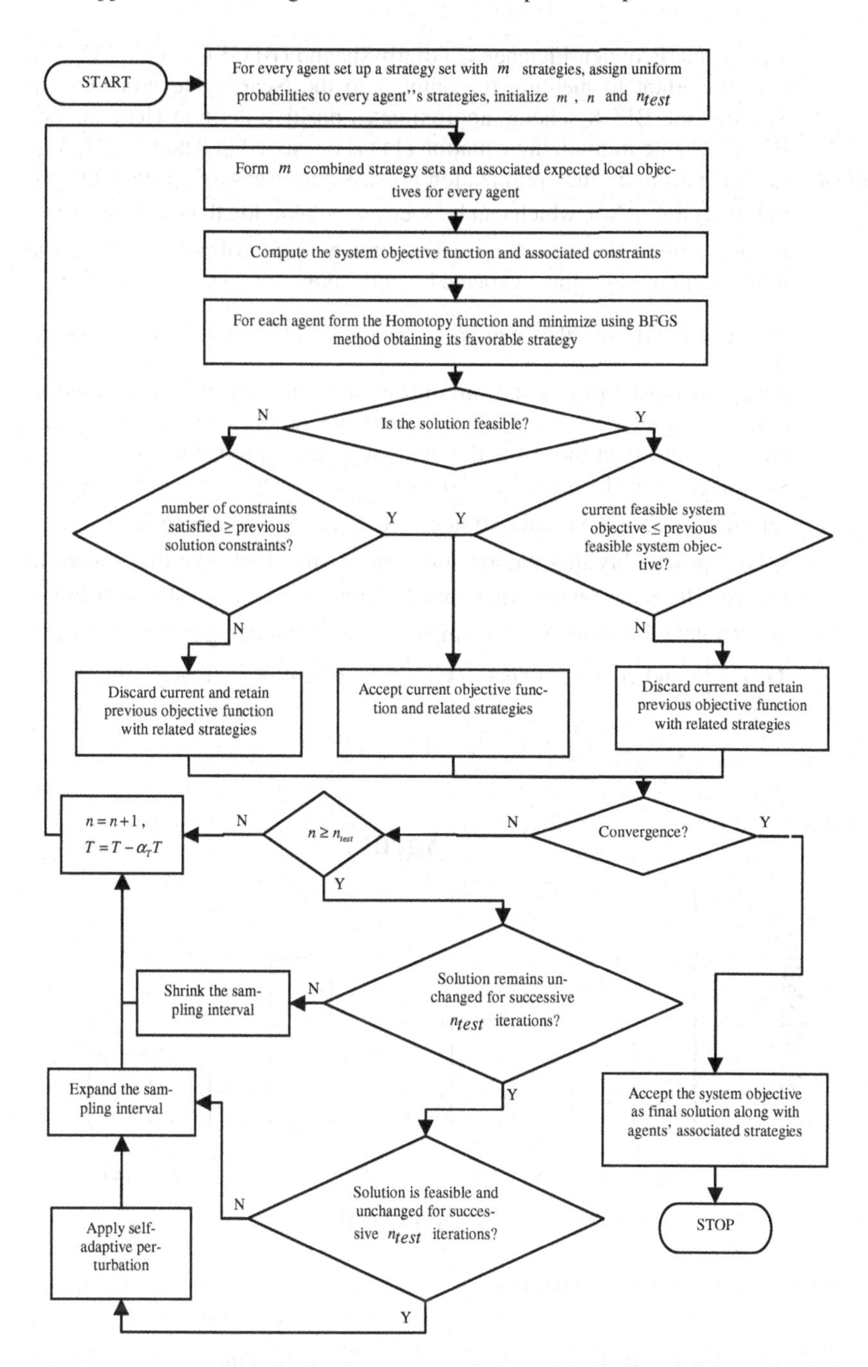

Fig. 1 Constrained PC Algorithm Flowchart

such as the Broyden-Fletcher-Goldfarb-Shanno (BFGS) scheme [32, 33]. It is important to mention that similar to the Nearest Newton Descent Scheme, the BFGS scheme approximates positive definite Hessian. The BFGS scheme minimizing equation (11) is discussed in detail in [32, 33].

Step 4. For each agent i, the optimization process converges to a probability variable vector $q(\mathbf{X}_i)$ which can be seen as the individual agent's probability distribution distinguishing every strategy's contribution towards the minimization of the expected collection of system objectives $\sum_{r=1}^{m_i} E\left(G(\mathbf{Y}_i^{[r]})\right)$. In other words, for every agent i, if strategy r contributes the most towards the minimization of the objective compared to other strategies, its corresponding probability certainly increases by some amount more than those for the other strategies' probability values, and so strategy r is distinguished from the other strategies. Such a strategy is referred to as a favorable strategy $X_i^{[fav]}$. As an illustration, the converged probability distribution for agent i may look like that shown in Figure 2 for a case where there are 10 strategies, i.e. $m_i = 10$ with favorable strategy indicated by *. Compute the corresponding system objective $G(\mathbf{Y}^{[fav]})$ and constraint vector $\mathbf{C}\left(\mathbf{Y}^{[fav]}\right)$ where $\mathbf{Y}^{[fav]}$ b is given by

$$\mathbf{Y}^{[fav]} = \left\{X_1^{[fav]}, X_2^{[fav]}, \ldots, X_{N-1}^{[fav]}, X_N^{[fav]}\right\} \tag{12}$$

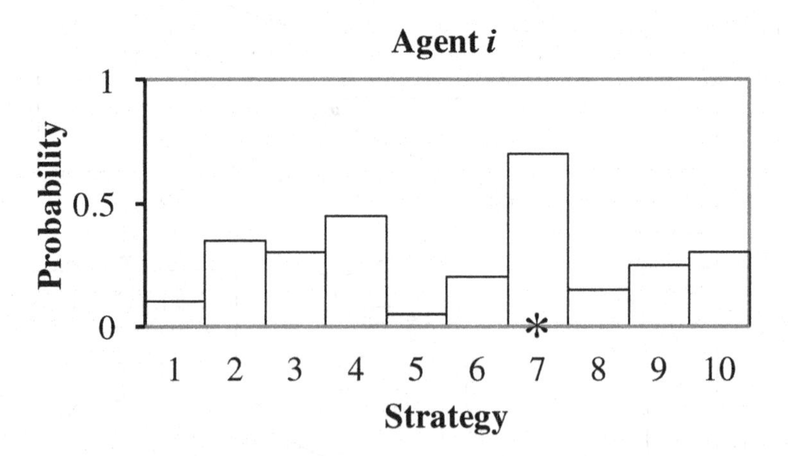

Fig. 2 Probability Distribution of agent i

Step 5. Accept the system objective $G(\mathbf{Y}^{[fav]})$ and corresponding $\mathbf{Y}^{[fav]}$ as current solution, if the number of constraints violated, $C_{violated} \le \mu$. Update the constraint violation tolerance $\mu = C_{violated}$ and continue to *step 6*.

If $C_{violated} > \mu$, discard current system objective $G(\mathbf{Y}^{[fav]})$ and corresponding $\mathbf{Y}^{[fav]}$, and retain the previous iteration solution and continue to *step 6*.

If the current system objective $G(\mathbf{Y}^{[fav]})$ is feasible i.e. $\mu = C_{violated} = 0$ and is not worse than the previous feasible solution, accept the current system objective $G(\mathbf{Y}^{[fav]})$ and corresponding $\mathbf{Y}^{[fav]}$ as current solution and continue to *step 6*, else discard current feasible system objective $G(\mathbf{Y}^{[fav]})$ and corresponding $\mathbf{Y}^{[fav]}$, and retain the previous iteration feasible solution and continue to *step 6*.

Step 6. On the completion of pre-specified n_{test} iterations, the following conditions are checked for every further iteration.

(a) If $G(\mathbf{Y}^{[fav],n}) \leq G(\mathbf{Y}^{[fav],n-n_{test}})$, then every agent shrinks its sampling interval as follows:

$$\Psi_i \in \left[\left(X_i^{[fav]} - \left\|\Psi_i^{upper} - \Psi_i^{lower}\right\| \cdot \lambda_{down}\right), \left(X_i^{[fav]} + \left\|\Psi_i^{upper} - \Psi_i^{lower}\right\| \cdot \lambda_{down}\right)\right]$$

, $0 < \lambda_{down} \leq 1$

where λ_{down} is referred to as the interval factor corresponding to the shrinking of sample space.

(b) If $G(\mathbf{Y}^{[fav],n})$ and $G(\mathbf{Y}^{[fav],n-n_{test}})$ are feasible and $\left\|G(\mathbf{Y}^{[fav],n}) - G(\mathbf{Y}^{[fav],n-n_{test}})\right\| \leq \varepsilon$, the system

objective $G(\mathbf{Y}^{[fav],n})$ can be referred to as a stable solution $G(\mathbf{Y}^{[fav],s})$ or possible local minimum. In order to jump out of this possible local minimum, a perturbation approach is incorporated. It is described below.

Every agent i perturbs its current favorable strategy $X_i^{[fav]}$ by a perturbation factor $fact_i$ corresponding to the reciprocal of its favorable strategy $X_i^{[fav]}$ as follows:

$$X_i^{[fav]} = X_i^{[fav]} \pm \left(X_i^{[fav]} \times fact_i\right) \tag{13}$$

where $fact_i = \begin{cases} random\,value \in \left(\sigma_1^{lower}, \sigma_1^{upper}\right) & if \ \dfrac{1}{X_i^{[fav]}} \leq \gamma \\ random\,value \in \left(\sigma_2^{lower}, \sigma_2^{upper}\right) & if \ \dfrac{1}{X_i^{[fav]}} > \gamma \end{cases}$

and $\sigma_1^{lower}, \sigma_1^{upper}, \sigma_2^{lower}, \sigma_2^{upper}$ are randomly generated values between 0 and 1 i.e. $0 < \sigma_1^{lower}, \sigma_1^{upper}, \sigma_2^{lower}, \sigma_2^{upper} < 1$, and

$\sigma_1^{lower} < \sigma_1^{upper} \leq \sigma_2^{lower} < \sigma_2^{upper}$. The value of γ as well as '+' or '-' sign in equation (13) are chosen based on the preliminary trials of the algorithm.

It gives a chance to every agent i to jump out of the local minima and further may help to search for a better solution. The perturbed solution is accepted if and only if the feasibility is maintained. Furthermore, every agent expands its sampling interval as follows:

$$\Psi_i \in \left[\left(\Psi_i^{lower} - \left\| \Psi_i^{upper} - \Psi_i^{lower} \right\| \cdot \lambda_{up} \right), \left(\Psi_i^{upper} + \left\| \Psi_i^{upper} - \Psi_i^{lower} \right\| \cdot \lambda_{up} \right) \right]$$

$$0 < \lambda_{up} \leq 1$$

where λ_{up} is referred to as the interval factor corresponding to the expansion of sample space.

Step 7. If either of the two criteria listed below is valid, accept the current stable system objective $G(\mathbf{Y}^{[fav],s})$ and corresponding $\mathbf{Y}^{[fav],s}$ as the final solution referred to as $G(\mathbf{Y}^{[fav],final})$ and $\mathbf{Y}^{[fav],final} = \left\{ X_1^{[fav],final}, X_2^{[fav],final}, \ldots, X_{N-1}^{[fav],final}, X_N^{[fav],final} \right\}$, respectively and stop, else continue to *step 8.*

(a) If temperature $T = T_{final}$ or $T \to 0$.

(b) If there is no significant change in the successive stable system objectives (i.e. $\left\| G(\mathbf{Y}^{[fav],s}) - G(\mathbf{Y}^{[fav],s-1}) \right\| \leq \varepsilon$) for two successive implementations of the perturbation approach.

Step 8. Each agent i then samples m_i strategies from within the updated sampling interval Ψ_i and forms the corresponding updated strategy set $\mathbf{X}_i$ represented as follows.

$$\mathbf{X}_i = \{X_i^{[1]}, X_i^{[2]}, X_i^{[3]}, \ldots, X_i^{[m_i]}\} \quad , \quad i = 1, 2, \ldots, N \tag{14}$$

Reduce the temperature $T = T - \alpha_T T$, update the iteration counter $n = n + 1$ and return to *step 1.*

From above detailed PC procedure, it is clear that in each iteration, every agent sets up its own strategy set/sampling interval and models the strategy set of every other agent based on their recent actions or behavior only, i.e. based on partial knowledge. In addition, each agent randomly samples the strategies from within its own strategy set as well as from within other agents' strategy sets, and computes the corresponding system objectives. By minimizing the collection of system objectives, every agent further identifies the possible strategy which contributes the most towards the minimization of the collection of system objectives. Hence, in every iteration of PC procedure, convergence of all the agents to the same solution cannot be guaranteed. However, the solutions achieved by all the agents could be guaranteed to be in very close proximity of one another.

2.2 Nash Equilibrium

In order to achieve a Nash equilibrium, every agent in a MAS should have the properties of Rationality and Convergence [49-52]. Rationality refers to the characteristic whereby every agent selects (or converges to) the best possible strategy given the strategies of the other agents. The convergence property refers to the stability condition i.e. a policy using which every agent selects (or converges to) the best possible strategy when all the other agents use their policies from a predefined class (preferably same class). The Nash equilibrium is naturally achieved when all the agents in a MAS are convergent and rational. Moreover, a Nash equilibrium is guaranteed when all the agents use stationary policies, i.e. those policies that do not change over time. It is worth to mention here that all the agents in the MAS proposed using PC algorithm exhibit the above mentioned properties. It is elaborated in the detailed PC algorithm discussed in the previous few paragraphs.

In any game, there may be a large but finite number of Nash equilibria present, depending on the number of strategies per agent as well as the number of agents. It is essential to choose the best possible combination of the individual strategies selected by each agent. It is quite hard to go through every possible combination of the individual agent strategies and choose the best out of it that can produce a best possible Nash equilibrium and hence the system objective.

As discussed in the detailed PC algorithm, in each iteration n, every agent i selects the best possible strategy referred to as the favorable strategy $X_i^{[fav],n}$ by guessing the possible strategies of the other agents. This information about its favorable strategy $X_i^{[fav],n}$ is made known to all the other agents as well. In addition, the corresponding global knowledge such as system objective value $G(\mathbf{Y}^{[fav],n}) = G\left(X_1^{[fav],n}, X_2^{[fav],n}, X_3^{[fav],n}, \ldots, X_{N-1}^{[fav],n}, X_N^{[fav],n}\right)$ is also available to each agent which clearly helps all the agents take the best possible informed decision in every further iteration. This makes the entire system ignore a considerably large number of Nash equilibria but select the best possible one in each iteration and accept the corresponding system objective $G(\mathbf{Y}^{[fav],n})$. Mathematically the Nash equilibrium solution in any iteration can be represented as follows:

$$
\begin{gathered}
G\left(X_1^{[fav],n}, X_2^{[fav],n}, \ldots, X_{N-1}^{[fav],n}, X_N^{[fav],n}\right) \le G\left(X_1^{(fav),n}, X_2^{[fav],n}, \ldots, X_{N-1}^{[fav],n}, X_N^{[fav],n}\right) \\
G\left(X_1^{[fav],n}, X_2^{[fav],n}, \ldots, X_{N-1}^{[fav],n}, X_N^{[fav],n}\right) \le G\left(X_1^{[fav],n}, X_2^{(fav),n}, \ldots, X_{N-1}^{[fav],n}, X_N^{[fav],n}\right) \\
\vdots \\
G\left(X_1^{[fav],n}, X_2^{[fav],n}, \ldots, X_{N-1}^{[fav],n}, X_N^{[fav],n}\right) \le G\left(X_1^{[fav],n}, X_2^{[fav],n}, \ldots, X_{N-1}^{[fav],n}, X_N^{(fav),n}\right)
\end{gathered}
\tag{15}
$$

where $X_i^{(fav),n}$ represents any strategy other than the favorable strategy $X_i^{[fav],n}$ from the same sample space Ψ_i^n.

Furthermore, from this current Nash equilibrium point with system objective $G(\mathbf{Y}^{[fav],n})$, the algorithm progresses to the next Nash equilibrium point with better system objective $G(\mathbf{Y}^{[fav],n+1})$, i.e. $G(\mathbf{Y}^{[fav],n}) \geq G(\mathbf{Y}^{[fav],n+1})$. As the algorithm progresses, those ignored Nash equilibria as well as the best Nash equilibria selected at previous iterations would be noticed as inferior solutions.

This process continues until there is no change in the current solution $G(\mathbf{Y}^{[fav],n})$, i.e. no new Nash equilibrium has been identified that proves the current Nash equilibrium to be inferior. Hence the system exhibits stage-wise convergence to a unique Nash equilibrium and the corresponding system objective is accepted as the final solution $G(\mathbf{Y}^{[fav],final})$. As a general case, this progress can be represented as

$$G(\mathbf{Y}^{[fav],1}) \geq G(\mathbf{Y}^{[fav],2}) \geq \ldots \geq G(\mathbf{Y}^{[fav],n}) \geq G(\mathbf{Y}^{[fav],n+1}) \geq \ldots \geq G(\mathbf{Y}^{[fav],final}).$$

3 The Circle Packing Problem (CPP)

The solution to a general packing problem aims to determine how best to pack z objects into a predefined bounded space that yields best utilization of space with no overlap of object boundaries [53, 54]. The bounded space can also be referred to as a container. The packing objects and container can be circular, rectangular or irregular. Although the problem appears rather simple and in spite of its practical applications in production and packing for the textile, apparel, naval, automobile, aerospace, food industries, etc. [55] the CPP received considerable attention in the 'pure' mathematics literature but only limited attention in the operations research literature [56]. According to [54, 57-59], as CPP cannot be effectively solved by purely analytical approaches [60-70], a number of heuristic techniques were proposed [53, 54, 71-83]. Most of these approaches address the CPP in limited ways, such as close packing of fixed and uniform sized circles inside a square or circle container [54, 60-69], close packing of fixed and different sized circles inside a square or circle container [76-83], simultaneous increase in the size of the circles covering the maximum possible area inside a square [72-75], etc. PC was successfully applied solving two cases of the CPP [32]. It is important to mention here that true optimum solution was achieved in every run of the PC solving both the cases of the CPP. The next few sections describe mathematical formulation of the CPP followed by the solution to the CPP without agent failure [32] and the case of the CPP with agent failure.

3.1 Formulation of the CPP

The objective of the CPP solved here was to cover the maximum possible area within a square by z number of circles without overlapping one another or exceeding the boundaries of the square. In order to achieve this objective, all the circles were allowed to increase their sizes as well as change their locations. The problem is formulated as follows:

$$\text{Minimize} \quad f = L^2 - \sum_{i=1}^{z} \pi r_i^2 \tag{16}$$

Subject to

$$\sqrt{\left(x_i - x_j\right)^2 + \left(y_i - y_j\right)^2} \geq r_i + r_j \tag{17}$$

$$x_i - r_i \geq x_l \tag{18}$$

$$x_i + r_i \leq x_u \tag{19}$$

$$y_i - r_i \geq y_l \tag{20}$$

$$y_i + r_i \leq y_u \tag{21}$$

$$0.001 \leq r_i \leq L/2 \tag{22}$$

$$i, j = 1, 2, \ldots, z \quad i \neq j \tag{23}$$

where

L = length of the side of the square
r_i = radius of circle i
x_i, y_i = x and y coordinates of the center of circle i
x_l, y_l = x and y coordinates of the lower left corner of the square
x_u, y_u = x and y coordinates of the upper right corner of the square

In solving the proposed CPP using constrained PC approach presented in Section 2.1, the circles were considered as autonomous agents. These circles were assigned the strategy sets of x and y coordinates of the center and the radius. In both cases of the CPP solved here, the circles were randomly initialized inside the square and were not allowed to cross the square boundaries. Moreover, for the case of the CPP with agent failure, on completion of a predefined number of iterations one of the circles was failed. Furthermore, for both the cases, the constraints in equation (17) were satisfied using the Feasibility-based Rule I described in Section 2.1 and the constraints in equations (18) to (21) were satisfied in every iteration of the algorithm using a repair approach. The repair approach refers to pushing the circles inside the square if they crossed the boundaries of it. It is similar to the one proposed in [1] solving the MTSP. The initial configuration of the case of the CPP without agent failure and the case with agent failure are shown in Figure 3(a) and 6(a), respectively.

The constrained PC algorithm solving both the cases was coded in MATLAB 7.8.0 (R2009A) and the simulations were run on a Windows platform using an Intel Core 2 Duo, 3GHz processor speed and 3.25GB memory capacity. Furthermore, for both the cases the set of parameters chosen was as follows: (a) individual agent sample size $m_i = 5$, (b) number of test iterations $n_{test} = 20$, (c) the shrinking interval factor $\lambda_{down} = 0.05$, (d) the expansion interval factor $\lambda_{up} = 0.1$, (e) perturbation parameters $\sigma_1^{lower} = 0.001$, $\sigma_1^{upper} = 0.01$, $\sigma_2^{lower} = 0.5$, $\sigma_2^{upper} = 0.7$, $\gamma = 0.99$ and the sign in equation (13) was chosen to be '-'. In addition to it, a voting heuristic was also incorporated in the constrained PC algorithm. It is described in Section 4.3.

3.2 CPP without Agent Failure

In this case of the CPP, five circles ($z = 5$) were initialized randomly inside the square and were not allowed to cross the square boundaries. The length of the side of the square was five units (i.e. $L = 5$). More than 30 runs of the constrained PC algorithm described in Section 2.1 with different initial configuration of the circles were conducted solving the CPP. The true optimum solution was achieved in every run with the average CPU time of 9 minutes and 25500 average number of function evaluations. The randomly generated initial solution, the intermediate iteration solutions, and the converged true optimum solution from one of the instances are presented in Figure 3. The corresponding convergence plot of the system objective is presented in Figure 4. The solution was converged at iteration 936 with 24336 function evaluations and the periodic rises represent the perturbations. The true optimum value of the objective function (f) achieved was 3.0807 units.

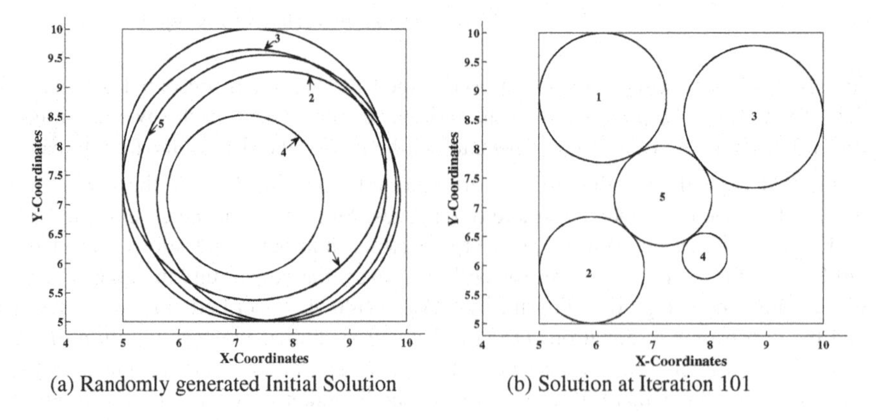

(a) Randomly generated Initial Solution (b) Solution at Iteration 101

Fig. 3 Solution History for CPP without Agent Failure

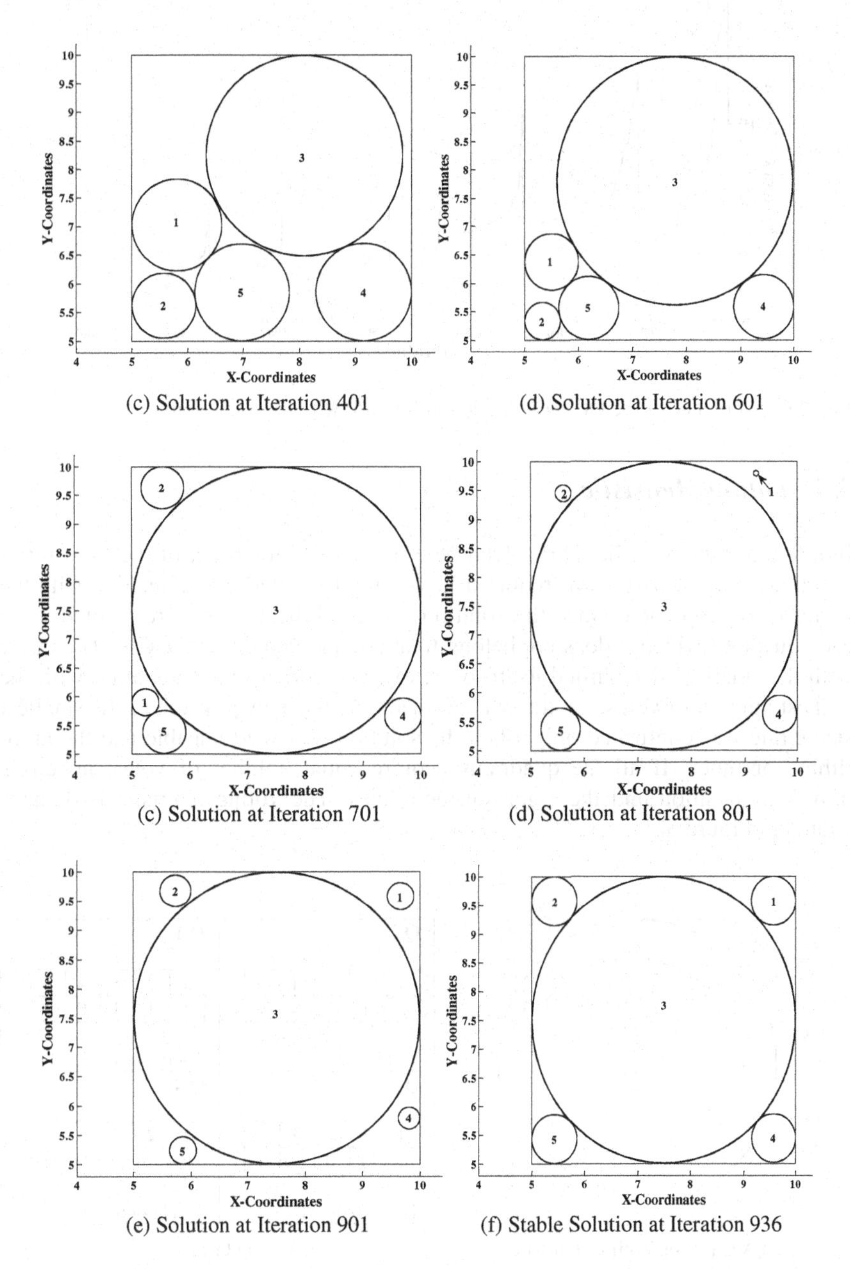

(c) Solution at Iteration 401

(d) Solution at Iteration 601

(c) Solution at Iteration 701

(d) Solution at Iteration 801

(e) Solution at Iteration 901

(f) Stable Solution at Iteration 936

Fig. 3 *(continued)*

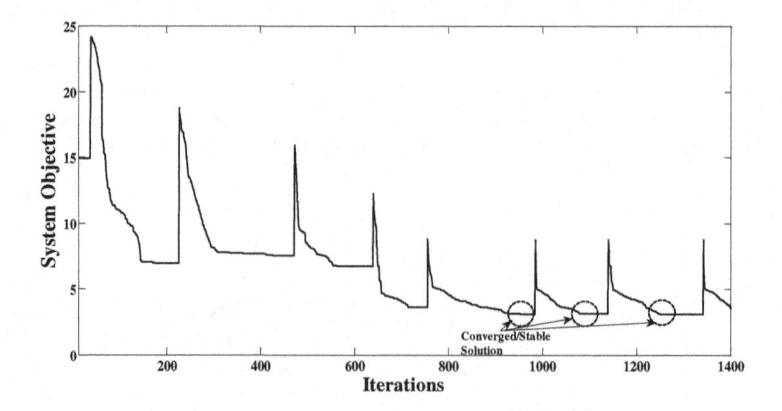

Fig. 4 Convergence of the Objective Function for CPP without Agent Failure

3.3 *Voting Heuristic*

In a few instances of the CPP solved here, in order to jump out of the local minimum, a voting heuristic was required. It was implemented in conjunction with the perturbation approach. Once the solution was perturbed, every circle voted 1 for each quadrant which it does not belong to at all, and voted 0 otherwise. The circle with the smallest size shifted itself to the extreme corner of the quadrant with the highest number of votes, i.e. the winner quadrant. The new position of the smallest size circle was confirmed only when the solution remained feasible and the algorithm continues. If all the quadrants acquire equal number of votes, no circle moves its position and the algorithm continues. The voting heuristic is demonstrated in Figure 5.

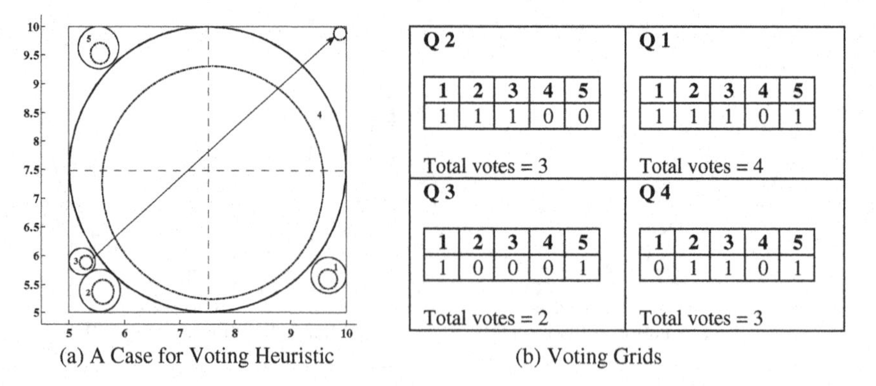

(a) A Case for Voting Heuristic (b) Voting Grids

Fig. 5 Voting Heuristic

A voting grid corresponding to every quadrant of the square in Figure 5(a) is represented in Figure 5(b). The solid circles represent the solution before perturbation while corresponding perturbed ones are represented in dotted lines. The votes given by the perturbed circles (dotted circles) to the quadrants are presented in the grid. As the maximum number of votes are given to quadrant 1 (i.e. Q 1), the circle with smallest size (circle 3) shifts to the extreme corner of the quadrant Q 1 and confirms the new position as the solution remains feasible. Based on the trials conducted so far, it was noticed that the voting heuristic was not necessary to be implemented in every run of the constrained PC algorithm solving the CPP. Moreover, in those of the few runs in which the voting heuristic was required, it was required to be implemented only once in the entire execution of the algorithm. A variant of the voting heuristic was also implemented in conjunction with energy landscape paving algorithm [53, 77, 78], in which the smallest circle was picked and placed randomly at the vacant place to produce a new configuration. It was claimed that this heuristic helped the algorithm jump out of the local minima. Furthermore, this heuristic was required to be implemented in every iteration of the algorithm.

3.4 CPP with Agent Failure

As mentioned in Section 3, the failed agent in PC can be considered as the one that does not communicate with other agents in the system and does not update its probability distribution. This does not prevent other agents from continuing further, by simply considering the failed agent's latest communicated strategies as the current strategies. In the context of the CPP, circle 2 was failed at a randomly chosen iteration 30, i.e. circle 2 does not update its x and y coordinates as well as its radius after iteration 30. More than 30 cases of the constrained PC algorithm with Feasibility-based Rule I were conducted solving the CPP with agent failure case and the average function evaluations were 17365. For the case presented here, the number of function evaluations was 19066 and the randomly generated initial solution, the intermediate iteration solutions, and the converged true optimum solution are shown in Figure 6. The corresponding convergence plot of the system objective is presented in Figure 7. The solution was converged at iteration 890 with 23140 function evaluations and the periodic rises represent the perturbations described in Section 2.1.

It is worth to mention that the voting heuristic was not required for the above agent failure case. This is because according to the voting heuristic the smallest circle that has to move to the new position will have to be the failed agent itself. In addition, the same set of parameter values listed in Section 4.1 was used for the agent failure case.

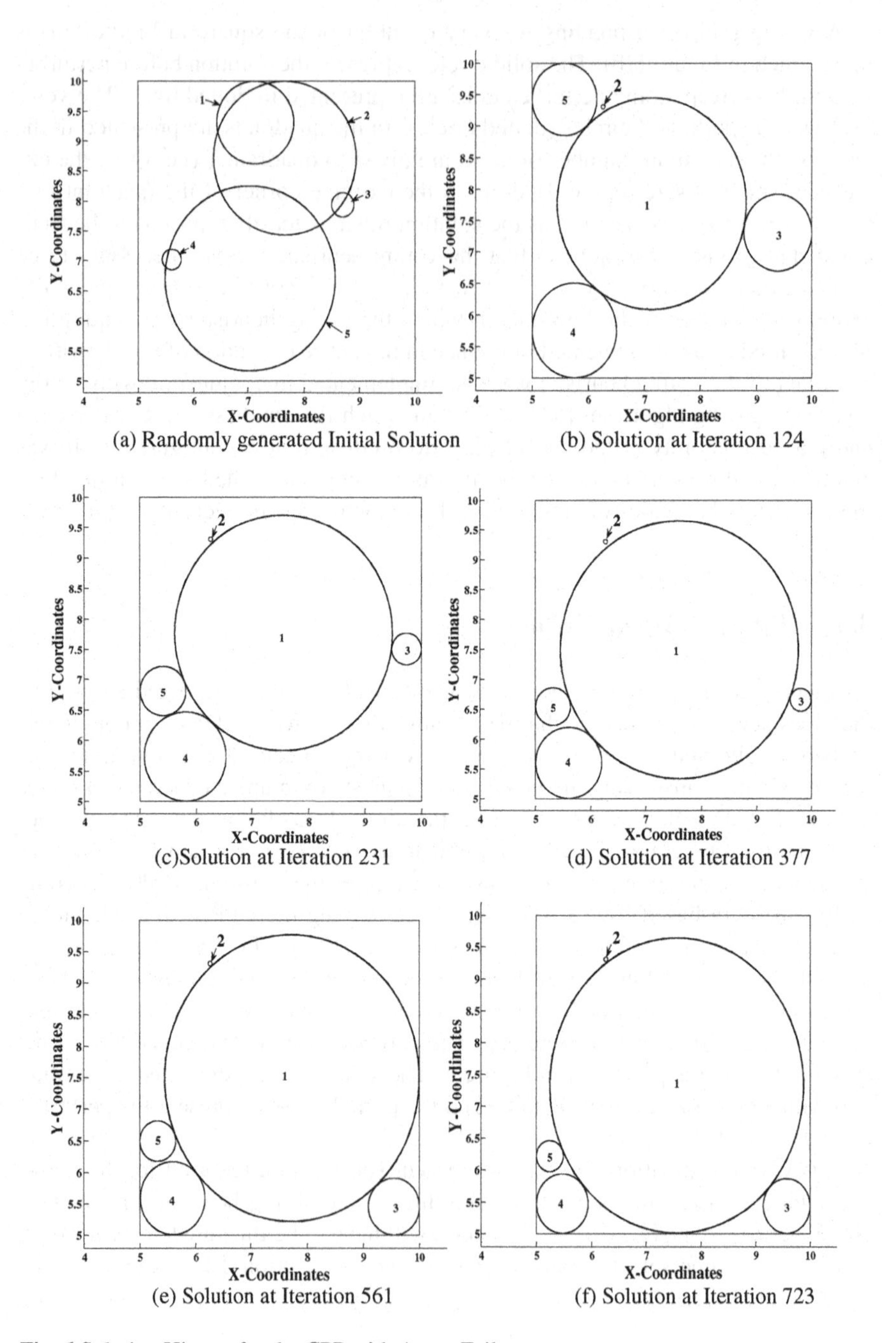

Fig. 6 Solution History for the CPP with Agent Failure

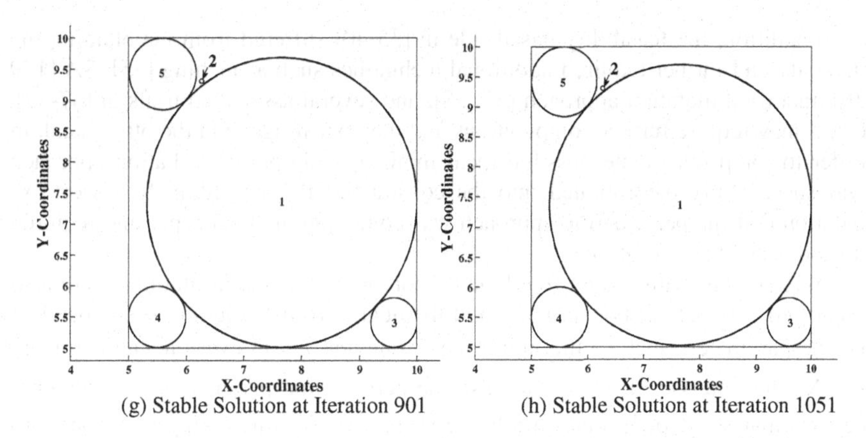

(g) Stable Solution at Iteration 901

(h) Stable Solution at Iteration 1051

Fig. 6 *(continued)*

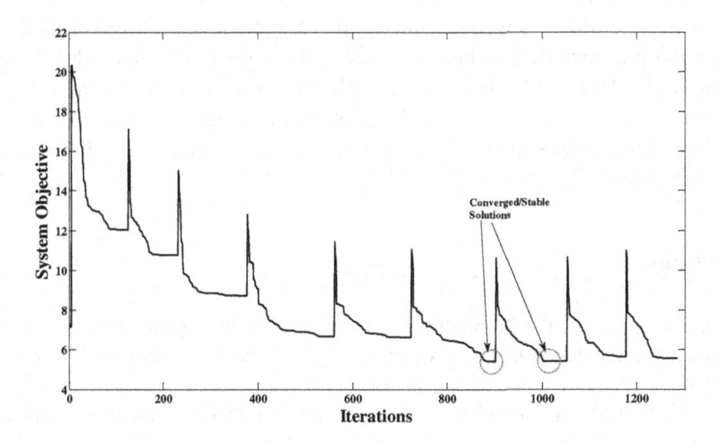

Fig. 7 Convergence of the Objective Function for the CPP with Agent Failure

4 Discussion and Conclusions

The above sections described the successful implementation of a generalized constrained PC approach using a variation of the feasibility-based rule originally proposed in [34]. It is evident that because of the inherent distributed nature of the PC algorithm, it can easily accommodate the agent failure case. The solution highlights its strong potential to deal with the agent failure which may arise in real world complex problems including urban traffic control, formation of airplanes fleet and mid-air collision avoidance, etc. In addition, it is evident that the approach was sufficiently robust and produced true optimum results in every run of the case of CPP without agent failure. It implies that the rational behavior of the agents could be successfully formulated and demonstrated. It is important to mention that the concept of the avoidance of tragedy of commons was also successfully demonstrated in solving the CPP.

In addition, the feasibility-based rule in [35-40] suffered from maintaining the diversity and further required additional techniques such as niching [35], SA [179 36], modified mutation approach [37, 38], and several associated trials in [37-39], etc. It may require further computations and memory usage. On the other hand, in order to jump out of the possible local minima, a simple perturbation approach was successfully incorporated into the constrained PC algorithm. It is worth to mention that the perturbation approach was computationally cheaper and required no additional memory usage.

In agreement with the no-free-lunch theorem [84], some limitations were also identified. The rate of convergence and the quality of the solution was dependent on the parameters such as the number of strategies m_i in every agent's strategy set $\mathbf{X}_i$, the interval factor λ and also the perturbation parameters. It is also necessary that some preliminary trials are performed for fine-tuning these parameters. Additionally, in order to confirm the convergence, the algorithm was required to be run beyond the convergence for a considerable number of iterations. In the future, as a generalized approach, an adaptive system can be developed for self tuning of the parameters. Furthermore, the ability of the PC methodology accommodating multiple agent failure case at multiple instances can be practically tested solving more realistic problems such as machine shop scheduling, urban traffic control, etc. The authors also see some potential in the field of healthcare systems management [85].

References

1. Kulkarni, A.J., Tai, K.: Probability Collectives: A Multi-Agent Approach for Solving Combinatorial Optimization Problems. Applied Soft Computing 10(3), 759–771 (2010)
2. Wolpert, D.H.: Information Theory – The Bridge Connecting Bounded Rational Game Theory and Statistical Physics. In: Braha, D., Minai, A.A., Bar-Yam, Y. (eds.) Complex Engineered Systems, pp. 262–290. Springer (2006)
3. Bieniawski, S.R.: Distributed Optimization and Flight Control Using Collectives. PhD Dissertation, Stanford University, CA, USA (2005)
4. Kulkarni, A.J., Tai, K.: Probability Collectives for Decentralized, Distributed Optimization: A Collective Intelligence Approach. In: Proceedings of the IEEE International Conference on Systems, Man, and Cybernetics, pp. 1271–1275 (2008)
5. Kulkarni, A.J., Tai, K.: Probability Collectives: A Decentralized, Distributed Optimization for Multi-Agent Systems. In: Mehnen, J., Koeppen, M., Saad, A., Tiwari, A. (eds.) Applications of Soft Computing, pp. 441–450. Springer (2009)
6. Prothmann, H., Rochner, F., Tomforde, S., Branke, J., Müller-Schloer, C., Schmeck, H.: Organic Control of Traffic Lights. In: Rong, C., Jaatun, M.G., Sandnes, F.E., Yang, L.T., Ma, J. (eds.) ATC 2008. LNCS, vol. 5060, pp. 219–233. Springer, Heidelberg (2008)
7. Akcelik, R.: Trafic Signals: Capacity and Timing Analysis, Australian Road Research Board (ARRB) Research Report, 123 (1981)

8. Stevanovic, A., Martin, P.T., Stevanovic, J.: VISGAOST: VISSIM-based Genetic Algorithm Optimization of Signal Timings. In: Proceedings of the 86th Transportation Reaseach Board Meeting (2007)
9. Branke, J., Goldate, P., Prothmann, H.: Actuated Traffic Signal Optimization using Evolutionary Algorithms. In: Proceedings of the 6th European Congress and Exhibition on Intelligent Transport Systems and Services (2007)
10. Prothmann, H., Branke, J., Schmeck, H., Tomforde, S., Rochner, F., Hahner, J., Muller-Schloer, C.: International Journal of Autonomous and Adaptive Communications Systems 2(3), 203–225 (2009)
11. Bazzan, A.L.C.: A Distributed Approach for Coordination of Traffic Signal Agents. Autonomous Agents and Multi-Agent Systems 10, 131–164 (2005)
12. Bhadra, S., Shakkotai, S., Gupta, P.: Min-cost selfish multicast with network coding. IEEE Transactions on Information Theory 52(11), 5077–5087 (2006)
13. Xi, Y., Yeh, E.M.: Distributed algorithms for minimum cost multicast with network coding in wireless networks. In: Proceedings of 4th International Symposium on Modeling and Optimization in Mobile, Ad Hoc and Wireless Networks (2006)
14. Yuan, J., Li, Z., Yu, W., Li, B.: A cross-layer optimization framework for multicast in multi-hop wireless networks. In: First International Conference on Wireless Internet, pp. 47–54 (2005)
15. Chatterjee, M., Sas, S.K., Turgut, D.: An on-demand weighted clustering algorithm (WCA) for ad hoc networks. In: Proceedings of 43rd IEEE Global Telecommunications Conference, pp. 1697–1701 (2000)
16. Amerimehr, M.H., Khalaj, B.K., Crespo, P.M.: A Distributed Cross-Layer OptimizationMethod for Multicast in Interference-Limited Multihop Wireless Networks. Journal onWireless Communications and Networking, Article ID 702036 (2008)
17. Mohammad, H.A., Babak, H.K.: A Distributed Probability Collectives Optimization Method for Multicast in CDMA Wireless Data Networks. In: Proceedings of 4th IEEE Internatilonal Symposium on Wireless Communication Systems, art. No. 4392414, pp. 617–621 (2007)
18. Ryder, G.S., Ross, K.G.: A probability collectives approach to weighted clustering algorithms for ad hoc networks. In: Proceedings of Third IASTED International Conference on Communications and Computer Networks, pp. 94–99 (2005)
19. Wolpert, D.H., Tumer, K.: An introduction to Collective Intelligence. Technical Report, NASA ARC-IC-99-63, NASA Ames Research Center (1999)
20. Wolpert, D.H., Strauss, C.E.M., Rajnarayan, D.: Advances in Distributed Optimization using Probability Collectives. Advances in Complex Systems 9(4), 383–436 (2006)
21. Basar, T., Olsder, G.J.: Dynamic Non-Cooperative Game Theory. Academic Press, London (1995)
22. Vasirani, M., Ossowski, S.: Collective-Based Multiagent Coordination: A Case Study. In: Artikis, A., O'Hare, G.M.P., Stathis, K., Vouros, G.A. (eds.) ESAW 2007. LNCS (LNAI), vol. 4995, pp. 240–253. Springer, Heidelberg (2008)
23. Huang, C.F., Chang, B.R.: Probability Collectives Multi-agent Systems: A Study of Robustness in Search. In: Pan, J.-S., Chen, S.-M., Nguyen, N.T. (eds.) ICCCI 2010, Part II. LNCS, vol. 6422, pp. 334–343. Springer, Heidelberg (2010)
24. Huang, C.F., Bieniawski, S., Wolpert, D., Strauss, C.E.M.: A Comparative Study of Probability Collectives Based Multiagent Systems and Genetic Algorithms. In: Proceedings of the Conference on Genetic and Evolutionary Computation, pp. 751–752 (2005)

25. Smyrnakis, M., Leslie, D.S.: Sequentially Updated Probability Collectives. In: Proceedings of 48thIEEE Conference on Decision and Control and 28th Chinese Control Conference, pp. 5774–5779 (2009)
26. Sislak, D., Volf, P., Pechoucek, M., Suri, N.: Automated Conflict Resolution Utilizing Probability Collectives Optimizer. IEEE Transactions on Systems, Man and Cybernetics: Applications and Reviews 41(3), 365–375
27. Wolpert, D.H., Antoine, N.E., Bieniawski, S.R., Kroo, I.R.: Fleet Assignment Using Collective Intelligence. In: Proceedings of the 42nd AIAA Aerospace Science Meeting Exhibit (2004)
28. Bieniawski, S.R., Kroo, I.M., Wolpert, D.H.: Discrete, Continuous, and Constrained Optimization Using Collectives. In: 10th AIAA/ISSMO Multidisciplinary Analysis and Optimization Conference, vol. 5, pp. 3079–3087 (2004)
29. Kulkarni, A.J., Tai, K.: Probability Collectives: A Distributed Optimization Approach for Constrained Problems. In: Proceedings of IEEE World Congress on Computational Intelligence, pp. 3844–3851 (2010)
30. Kulkarni, A.J., Tai, K.: Solving Constrained Optimization Problems Using Probability Collectives and a Penalty Function Approach. International Journal of Computational Intelligence and Applications 10(4), 445–470 (2011)
31. Autry, B.: University Course Timetabling with Probability Collectives. Master's Thesis, Naval Postgraduate School Montery, CA, USA (2008)
32. Kulkarni, A.J., Tai, K.: A Probability Collectives Appraoch with a Feasibility-based Rule for Constrained Optimization. Applied Computational Intelligence and Soft Computing, Article ID 980216 (2011)
33. Kulkarni, A.J., Tai, K.: Solving Sensor Network Coverage Problems Using a Constrained Probability Collectives Approach. Submitted to Applied Soft Computing (July 2011)
34. Deb, K.: An Efficient Constraint Handling Method for Genetic Algorithms. Computer Methods in Applied Mechanics in Engineering 186, 311–338 (2000)
35. He, Q., Wang, L.: A Hybrid Particle Swarm Optimization with a Feasibility-based Rule for Constrained Optimization. Applied Mathematics and Computation 186, 1407–1422 (2007)
36. Sakthivel, V.P., Bhuvaneswari, R., Subramanian, S.: Design Optimization of Three-Phase Energy Efficient Induction Motor Using Adaptive bacterial Foraging Algorithm. Computation and Mathematics in Electrical and Electronic Engineering 29(3), 699–726 (2010)
37. Kaveh, A., Talatahari, S.: An Amproved Ant Colony Optimization for Constrained Engineering Design Problems. Computer Aided Engineering and Software 27(1), 155–182 (2010)
38. Bansal, S., Mani, A., Patvardhan, C.: Is Stochastic Ranking Really Better than Feasibility Rules for Constraint Handling in Evolutionary Algorithms? In: Proceedings of World Congress on Nature and Biologically Inspired Computing, pp. 1564–1567 (2009)
39. Gao, J., Li, H., Jiao, Y.C.: Modified Differential Evolution for the Integer Programming Problems. In: Proceedings of International Conference on Artificial Intelligence, pp. 213–219 (2009)
40. Ping, W., Xuemin, T.: A Hybrid DE-SQP Algorithm with Switching Procedure for Dynamic Optimization. In: Proceedings of Joint 48th IEEE Conference on Decision and Control and 28th Chinese Control Conference, pp. 2254–2259 (2009)

41. Sindhwani, V., Keerthi, S.S., Chapelle, O.: Deterministic Annealing for Semi-supervised Kernel Machines. In: Proceedings of the 23th International Conference on Machine Learning (2006)
42. Rao, A.V., Miller, D.J., Rose, K., Gersho, A.: A Deterministic Annealing Approach for Parsimonious Design of Piecewise Regression Models. IEEE Transactions on Patternt Analysis and Machine Intelligence 21(2) (1999)
43. Rao, A.V., Miller, D.J., Rose, K., Gersho, A.: Mixture of Experts Regression Modeling by Deterministic Annealing. IEEE Transactions on Signal Processing 45(11) (1997)
44. Czabański, R.: Deterministic Annealing Integrated with ε-Insensitive Learning in Neuro-fuzzy Systems. In: Rutkowski, L., Tadeusiewicz, R., Zadeh, L.A., Żurada, J.M. (eds.) ICAISC 2006. LNCS (LNAI), vol. 4029, pp. 220–229. Springer, Heidelberg (2006)
45. `http://hyperphysics.phy-astr.gsu.edu/hbase/thermo/helmholtz.html` (last accessed on: December 12, 2011)
46. Ray, T., Tai, K., Seow, K.C.: An Evolutionary Algorithm for Constrained Optimization. In: Proceedings of the Genetic and Evolutionary Computation Conference, pp. 771–777 (2000)
47. Ray, T., Tai, K., Seow, K.C.: Multiobjective Design Optimization by an Evolutionary Algorithm. Engineering Optimization 33(4), 399–424 (2001)
48. Tai, K., Prasad, J.: Target-Matching Test Problem for Multiobjective Topology Optimization Using Genetic Algorithms. Structural and Multidisciplinary Optimization 34(4), 333–345 (2007)
49. Shoham, Y., Powers, R., Grenager, T.: Multi-agent reinforcement learning: A critical survey. Department of Computer Science, Stanford University CA, USA, Technical Report (2003), `http://multiagent.stanford.edu/papers/MALearning_ACriticalSurvey_2003_0516.pdf`
50. Busoniu, L., Babuska, L., Schutter, B.: A Comprehensive Survey of Multiagent Reinforcement Learning. IEEE Transactions on Systems, Man, and Cybernetics—Part C: Applications and Reviews 38(2), 156–172 (2008)
51. Bowling, M., Veloso, M.: Multiagent learning using a variable learning rate. Artificial Intelligence 136(2), 215–250 (2002)
52. Bowling, M., Veloso, M.: Rational and convergent learning in stochastic games. In: Proceedings of 17th International Conference on Artificial Intelligence, pp. 1021–1026 (2001)
53. Zhang, D., Deng, A.: An effective hybrid algorithm for the problem of packing circles into a larger containing circle. Computers & Operations Research 32, 1941–1951 (2005)
54. Theodoracatos, V.E., Grimsley, J.L.: The optimal packing of arbitrarily-shaped polygons using simulated annealing and polynomial-time cooling schedules. Computer Methods in Applied Mechanics and Engineering 125, 53–70 (1995)
55. Liu, J., Xue, S., Liu, Z., Xu, D.: An improved energy landscape paving algorithm for the problem of packing circles into a larger containing circle. Computers & Industrial Engineering 57(3), 1144–1149 (2009)
56. Castillo, I., Kampas, F.J., Pinter, J.D.: Solving Circle Packing Problems by Global Optimization: Numerical Results and Industrial Applications. European Journal of Operational Research 191(3), 786–802 (2008)

57. Garey, M.R., Johnson, D.S.: Computers and intractability: A Guide to the Theory of NP-completeness. W. H. Freeman & Co. (1979)
58. Hochbaum, D.S., Maass, W.: Approximation schemes for covering and packing problems in image processing and VLSI. Journal of the Association for Computing Machinery 1(32), 130–136 (1985)
59. Wang, H., Huang, W., Zhang, Q., Xu, D.: An improved algorithm for the packing of unequal circles within a larger containing circle. European Journal of Operational Research 141, 440–453 (2002)
60. Szabo, P.G., Markot, M.C., Csendes, T.: Global optimization in geometry - circle packing into the square. In: Audet, P., Hansen, P., Savard, G. (eds.) Essays and Surveys in Global Optimization, pp. 233–265. Kluwer (2005)
61. Nurmela, K.J., Ostergard, P.R.J.: Packing up to 50 equal circles in a square. Discrete & Computational Geometry 18, 111–120 (1997)
62. Nurmela, K.J., Ostergard, P.R.J.: More Optimal Packing of Equal Circles in a Square. Discrete and Computational Geometry 22, 439–457 (1999)
63. Graham, R.L., Lubachevsky, B.D.: Repeated Patterns of Dense Packings of Equal Disks in a Square. The Electronic Journal of Combinatorics 3, R16 (1996)
64. Szabo, P.G., Csendes, T., Casado, L.G., Garcia, I.: Equal circles packing in a square I - Problem Setting and Bounds for Optimal Solutions. In: Giannessi, F., Pardalos, P., Rapcsak, T. (eds.) Optimization Theory: Recent Developments from Matrahaza, pp. 191–206. Kluwer (2001)
65. de Groot, C., Peikert, R., Wurtz, D.: The Optimal Packing of Ten Equal Circles in a Square, IPS Research Report 90-12, ETH, Zurich (1990)
66. Goldberg, M.: The packing of equal circles in a square. Mathematics Magazine 43, 24–30 (1970)
67. Mollard, M., Payan, C.: Some Progress in the Packing of Equal Circles in a Square. Discrete Mathematics 84, 303–307 (1990)
68. Schaer, J.: On the Packing of Ten Equal Circles in a Square. Mathematics Magazine 44, 139–140 (1971)
69. Schluter, K.: Kreispackung in Quadraten. Elemente der Mathematics 34, 12–14 (1979)
70. Valette, G.: A Better Packing of Ten Circles in a Square. Discrete Mathematics 76, 57–59 (1989)
71. Boll, D.W., Donovan, J., Graham, R.L., Lubachevsky, B.D.: Improving Dense Packings of Equal Disks in a Square. The Electronic Journal of Combinatorics 7, R46 (2000)
72. Lubachevsky, D., Graham, R.L.: Curved hexagonal packing of equal circles in a circle. Discrete & Computational Geometry 18, 179–194 (1997)
73. Szabo, P.G., Specht, E.: Packing up to 200 Equal Circles in a Square. In: Torn, A., Zilinskas, J. (eds.) Models and Algorithms for Global Optimization, pp. 141–156 (2007)
74. Mladenovic, N., Plastria, F., Urosevi, D.: Formulation Space Search for Circle Packing Problems. LNCS, pp. 212–216 (2007)
75. Mldenovic, N., Plastria, F., Urosevi, D.: Reformulation Descent Applied to Circle Packing Problems. Computers and Operations Research 32, 2419–2434 (2005)
76. Huang, W., Chen, M.: Note on: An Improved Algorithm for the Packing of Unequal Circles within a Larger Containing Circle. Computers and Industrial Engineering 50(2), 338–344 (2006)
77. Liu, J., Xu, D., Yao, Y., Zheng, Y.: Energy Landscape Paving Algorithm for Solving Circles Packing Problem. In: International Conference on Computational Intelligence and Natural Computing, pp. 107–110 (2009)

78. Liu, J., Yao, Y., Zheng, Y., Geng, H., Zhou, G.: An Effective Hybrid Algorithm for the Circles and Spheres Packing Problems. In: Du, D.-Z., Hu, X., Pardalos, P.M. (eds.) COCOA 2009. LNCS, vol. 5573, pp. 135–144. Springer, Heidelberg (2009)
79. Stoyan, Y.G., Yaskov, G.N.: Mathematical Model and Solution Method of Optimization Problem of Placement of Rectangles and Circles Taking into Account Special Constraints. International Transactions in Operational Research 5, 45–57 (1998)
80. Stoyan, Y.G., Yaskov, G.N.: A Mathematical Model and a Solution Method for the Problem of Placing Various-Sized Circles into a Strip. European Journal of Operational Research 156, 590–600 (2004)
81. George, J.A.: Multiple Container Packing: A Case Study of Pipe Packing. Journal of the Operational Research Society 47, 1098–1109 (1996)
82. George, J.A., George, J.M., Lamar, B.W.: Packing Different-sized Circles into a Rectangular Container. European Journal of Operational Research 84, 693–712 (1995)
83. Hifi, M., M'Hallah, R.: Approximate Algorithms for Constrained Circular Cutting Problems. Computers and Operations Research 31, 675–694 (2004)
84. Wolpert, D.H., Macready, W.G.: No Free Lunch Theorems for Optimization. IEEE Transactions on Evolutionary Computation 1(1), 67–82 (1997)
85. Ng, E.Y.K., Tai, K., Ng, W.K., Acharya, R.U.: Optimization of the Pharmacokinetic Simulation Models for the Nanoparticles-in-Nanoshapes Hybrid Drug Delivery System Using Heat Diffusion Analogy. In: Rajendra Acharya, U., Tamura, T., Ng, E.Y.K., Lim, C.M., Suri, J.S. (eds.) Distributed Diagnosis and Home Healthcare, pp. 211–230. American Scientific Publishers (2010)

Author Index

Barbucha, Dariusz 55, 123
Boryczka, Mariusz 29
Bura, Wojciech 29

Czarnowski, Ireneusz 1, 123

Jędrzejowicz, Piotr 1, 77, 103, 123

Kulkarni, Anand J. 175

Nedić, Angelia 143

Ratajczak-Ropel, Ewa 103, 123

Skakovski, Aleksander 77
Srivastava, Kunal 143
Stipanović, Dušan 143

Tai, Kang 175

Wierzbowska, Izabela 123